ZOOLOGICAL CHARACTERISTICS AND BEHAVIOUR OF ANIMALS

ZOOLOGICAL CHARACTERISTICS AND BEHAVIOUR OF ANIMALS

R.P. Singh

RANDOM PUBLICATIONS

NEW DELHI - 110 002 (INDIA)

Zoological Characteristics and Behaviour of Animals

ISBN 978-93-51116-91-2

Published in 2015 in India by

RANDOM PUBLICATIONS

4376-A/4B, Gali Murari Lal, Ansari Road
New Delhi-110 002
Phone: +9111-43580356, 23289044
E-mail: randomexports@gmail.com; sales@randompublications.com; info@randompublications.com

Reprinted 2023

Type Setting by: Friends Media, Delhi-110089
Digitally Printed at: Replika Press Pvt. Ltd.

Preface

Ethology is the scientific and objective study of animal behaviour under natural conditions, as opposed to behaviourism, which focuses on behavioural response studies in a laboratory setting. Ethologists have been particularly concerned with the evolution of behaviour and the understanding of behaviour in terms of the theory of natural selection. The desire to understand animals has made ethology a rapidly growing field. Since the turn of the 21st century, many aspects of animal communication, animal emotions, animal culture, learning, and even sexual conduct that experts long thought they understood, have been re-examined, and new conclusions reached. New fields have developed, such as neuroethology. Understanding ethology or animal behaviour can be important in animal training. Considering the natural behaviours of different species or breeds enables the trainer to select the individuals best suited to perform the required task. It also enables the trainer to encourage the performance of naturally occurring behaviours and also the discontinuance of undesirable behaviours. For ethologists, instinct means a series of predictable behaviours for fixed action patterns. Such schemes are only acted when a precise stimulating signal is present. When such signals act as communication among members of the same species, they are known as releasers. A notable example of a releaser is the beak movements in many bird species performed by the newly hatched chicks, which stimulates the mother's regurgitating process to feed her offspring.

Habituation is a simple form of learning and occurs in many animal taxa. It is the process whereby an animal ceases responding to a stimulus. Often in social life, animals fight for the right to reproduce, as well as social supremacy. A common example of fighting for social and sexual supremacy is the so-called pecking order among poultry. Several animal species, including humans, tend to live in groups. Group size is a major aspect of their social environment. Social life is probably a complex and effective survival strategy. It may be regarded as a sort of symbiosis among individuals of the same species: a society is composed of a group of individuals belonging to the same species living within well-defined rules on food management, role assignments and reciprocal dependence. The study and interpretation of the natural behaviour of animals is a fascinating subject and those who choose to study animal behaviour are concerned with understanding the causes, functions and evolution of behaviour. There is an increasing need for humans to take an element of responsibility towards animals and their welfare. Not just for domestic pets but for wild animals, who have to adapt to new environments as a result of climate change and the effects that this is having on the natural habitats of some wild animals.

This is an essential source of reference for students and teachers of zoology as well as naturalists.

I thank all members of my team who have helped in the preparation of the book. My special thanks go to "Random Publications" who have published the book.

— R.P. Singh

Contents

Chapter 1

Introduction

The opening paragraphs depict a hypothetical scenario of man first taming the wolf. Although we have tried to make light of this event, the fact is that no one knows exactly how or why this first encounter took place. The earliest archeological estimate indicates that it occurred in the late Glacial period, approximately 14,000 years BC (Boessneck, 1985). Another scenario is that wolves domesticated themselves. The presumption is that calm wolves with low levels of fear were likely to scavenge near human settlements. Both Coppinger and Smith (1983) and Zeuner (1963) suggest that wild species which later became domesticants started out as camp followers. Some wolves were believed to have scavenged near human settlements or followed hunting parties; wild cattle supposedly invaded grain fields, and wild cats may have invaded grainaries while hunting for mice. However, the most recent evidence obtained by sequencing mitochondrial DNA of 67 dog breeds and wolves from 27 localities indicates that dogs may have diverged from wolves over 100,000 years ago.

In any event, wolves kept for companions had to be easy to handle and socialise to humans. Within a few generations, early humans may have turned wolves into dogs by selecting and breeding the tamest ones. Thousands of years ago, humans were not aware that behaviour in animals was heritable. However, even today people who raise dogs, horses, pigs, cattle, or chickens notice differences in the behaviour of the offspring. Some animals are friendly and readily approach people, while others may be shy and nervous.

Genetic Effects of Domestication

Price (1984) defined domestication as a process by which a population of animals becomes adapted to man and the captive environment by some combination of genetic changes occurring over generations and environmentally induced developmental events recurring during each generation:' In long-term selection experiments designed to study the consequences of selection for the tame" domesticated type of behaviour, Belyaev (1979) and Belyaev *et al.* (1981) studied foxes reared for their fur. The red fox (*Vulpes fulva*) has been raised on seminatural fur farms for over 100 years and was selected for fur traits and not behavioural traits.

However, they demonstrate three distinctly different characteristic responses to man. Thirty percent were extremely aggressive toward man, 60% were either fearful or fearfully aggressive, and 10% displayed a quiet exploratory reaction without either fear or aggression. The objective of this experiment was to breed animals similar in behaviour to the domestic dog. By selecting and breeding the tamest individuals, 20 years later the experiment succeeded in turning wild foxes into tame, border collie-like fox-dogs. The highly selected "tame" population of (fox-dog) foxes actively sought human contact and would whine and wag their tails when people approached (Belyaev 1979). This behaviour was in sharp contrast to wild foxes which showed extremely aggressive and fearful behaviour toward man. Keeler *et al.* (1970) described this behaviour:

> Vulpes fulva *(the wild fox) is a bundle of jangled nerves. We had observed that when first brought into captivity as an adult, the red fox displays a number of symptoms that are in many ways similar to those observed in psychosis.*

They resemble a wide variety of phobias, especially fear of open spaces, movement, white objects, sounds, eyes or lenses, large objects, and man, and they exhibit panic, anxiety, fear, apprehension and a deep trust of the environment. They are: 1) catalepsy-like frozen positions, accompanied by blank stares; 2) fear of sitting down; 3) withdrawal; 4) runaway flight reactions; and 5) aggressiveness. Sometimes the strain of captivity makes them deeply disturbed and confused, or may produce a depression- like state. Extreme excitation and restlessness may also be observed in some individuals in response to many changes in the physical environment. Most adult red foxes soon after capture break off their canine teeth on the mesh of our expanded metal cage in their attempts to escape. A newly captured fox is known to have torn at the wooden door of his cage in a frenzy until he dropped dead from exhaustion.

Although the stress of domestication is great, Belyaev (1979) and Belyaev *et al.* (1981) concluded that selection for tameness was effective in spite of the many undesirable characteristics associated with tameness. For example, the tame foxes shed during the wrong season and developed black and white patterned fur, and changes were found in their hormone profiles. This means that the monoestrus (once a year) cycle of reproduction was disturbed and the animals would breed at any time of the year. Furthermore, changes in behaviour occurred simultaneously with changes in tail position and ear shape, and the appearance of a white muzzle, forehead blaze, and white shoulder hair. The white colour pattern on the head is similar to many domestic animals (Belyaev 1979). The most dog-like foxes had white spots and patterns on their heads, drooping ears, and curled tails and looked more like dogs than the foxes that avoided people. The behavioural and morphological (appearance) changes were also correlated with corresponding changes in the levels of gender hormones. The tame foxes had higher levels of the neurotransmitter serotonin (Popova *et al.*, 1975). Serotonin is known to inhibit some kinds of aggression (Belyaev, 1979), and serotonin ~levels are increased in the brains of people who take Prozac (fluoxetine).

The study of behavioural genetics can help explain why selection for calm temperament was linked to physical and neurochemical changes in Belyaev's foxes. Behaviour geneticists and animal scientists are interested in understanding effects on behaviour due to genetic influences or those which are due to environment and learning.

History

This historical review is not intended to he comprehensive; our objective is to discuss some of the early discoveries that are important for our current understanding of animal behaviour, with particular emphasis on genetic influence on behaviour in domestic animals.

Early in the 17th century, Descartes came to the conclusion "that the bodies of animals and men act wholly like machines and move in accordance with purely mechanical laws" (in Huxley 1874). After Descartes, others undertook the task of explaining behaviour as reactions to purely physical, chemical, or mechanical events. For the next three centuries scientific thought on behaviour oscillated between a mechanistic view that animals are '~automatons" moving through life without consciousness or self-awareness and an opposing view that animals had thoughts and feelings similar to those of humans.

In "On the Origin of the Species" (1859), Darwin's ideas about evolution began to raise serious doubts about the mechanistic view of animal behaviour. He noticed that animals share many physical characteristics and was one of the first to discuss variation within a species, both in their behaviour and in their physical appearance. Darwin believed that artificial selection and natural selection were intimately associated (Darwin, 1868) and cleverly outlined the theory of evolution without any knowledge of genetics. In "The Descent of Man" (1871) Darwin concluded that temperament traits in domestic animals are inherited. He also believed, as did many other scientists of his tune, that animals have subjective sensations and could think. Darwin wrote: "The differences in mind between man and the higher animals, great as it is, is certainly one of degree and not of kind."

Other scientists realised the implications of Darwin's theory on animal behaviour and conducted experiments investigating instinct. Herrick (1908) observed the behaviour of wild birds in order to determine, first, how their instincts are modified by their ability' to learn, and second, the degree of intelligence they attain. On the issue of thinking in animals, Schroeder (1914) concluded: "The solution, if it ever comes, can scarcely fail to illuminate, if not the animal mind, at least that of man." It is evident that by the end of the 19th century, scientists who studied animal behaviour in natural environments learned that the mechanical approach could not explain all behaviour.

During the middle of the 20th century', scientific thought again reverted to the mechanical approach and behaviourism reigned throughout America. The behaviourists ignored both genetic effects on behaviour and the ability of animals to engage in flexible problem solving. The founder of behaviourism, J. B. Watson (1930), stated that differences in the environment can explain all differences in behaviour." He did not believe that genetics had any effect on behaviour. In The behaviour of Organisms]' the psychologist B. F Skinner (1958) wrote that all behaviour could be explained by the principles of stimulus-response and operant conditioning.

The first author visited with Dr. Skinner at Harvard University in 1968. Skinner responded to a question from her about the need for brain research by saying, "We don't need to know about the brain because we have operant conditioning" (T. Grandin, personal communication, 1968). Operant conditioning uses food rewards and punishments to train animals and shape their behaviour. In a simple Skinner box experiment, a rat can be trained to push a lever to obtain food when a green light turns on, or to push a lever very quickly to avoid a shock when a red light appears. The signal light is the '~conditioned stimulus." Rats and other animals can be trained to perform a complex sequence of behaviours by chaining together a series of simple operant responses. Skinner believed that even the most complex behaviours can be explained as a series of conditioned responses.

However, a rat's behaviour is very limited in a Skinner box. It's a world with very little variation, and the rat has little opportunity to use its natural behaviours. It simply learns to push a lever to obtain food or prevent a shock. Skinnerian principles explain why a rat behaves a certain way in the sterile confines of a 30 x 30 cm Plexiglas box, but they don't reveal much about the behaviour of a rat in the local dump. Outside of the laboratory, a rat's behaviour is more complex.

Instincts versus Learning

Skinner's influence on scientific thinking slowed a bit in 1961 following the publication of "The Misbehaviour of Organisms" by Brelands and Brelands. This paper described how Skinnerian behavioural principles collided with instincts. The Brelands were trained Skinnerian behaviourists who attempted to apply the strict principles of operant conditioning to animals trained at fairs and carnivals. Ten years before this classic paper, the Brelands (1951) wrote, we are wholly affirmative and optimistic that principles derived from the laboratory can be applied to the extensive control of animal behaviour under non laboratory condition]' However, by 1961, after training more than 6000 animals as diverse as reindeer, cockatoos, raccoons, porpoises, and whales for exhibition in zoos, natural history museums, department store displays, fair and trade convention exhibits, and television, the Brelands wrote a second article featured in the American Psychologist (1961), which stated, our backgrounds in behaviourism had not prepared us for the shock of some of our failures."

One of the failures occurred when the Brelands tried to teach chickens to stand quietly on a platform for 10 to 12 seconds before they received a food reward. The chickens would stand quietly on a platform in the beginning of training; however, once they learned to associate the platform with a food reward, half (50%) started scratching the platform, and another 25% developed other behaviours, such as pecking the platform. The Brelands salvaged this disaster by developing a wholly unplanned exhibit involving a chicken that turned on a juke box and danced. They first trained the chickens to pull a rubber loop which turned on some music. When the music started, the chickens would jump on the platform and start scratching and pecking until the food reward was delivered. This exhibit made use of the chicken's instinctive food-getting behaviour. The first author remembers as a teenager seeing a similar exhibit, at the Arizona State Fair, of a piano-playing chicken in a little red barn. The hen would peck the keys of a toy piano when a quarter was put in the slot and would stop when the food came down the chute. This exhibit also worked because it was similar to a Skinner box in the laboratory.

The Brelands experienced another classic failure when they tried to teach raccoons to put coins in a piggy bank. Because raccoons are adept at manipulating objects with their hands, this task was initially easy. As training progressed, however, the raccoons began to rub the coins before depositing them in the bank.

This behaviour was similar to the washing behaviour raccoons do as instinctive food-getting behaviour. The raccoons at first had difficulty letting go of the coin and would hold and rub it. However, when the Brelands introduced a second coin, the raccoons became almost impossible to train. Rubbing the coins together 'in a most miserly fashion]' the raccoons got worse and worse as time went on. The Brelands concluded that the innate behaviours were suppressed during the early stages of training and sometimes long into the training, but as training progressed, instinctive food-getting behaviours gradually replaced the conditioned behaviour. The animals were unable to override their instincts and thus a conflict between conditioned and instinctive behaviours occurred.

Ethology

While Skinner and his fellow Americans were refining the principles of operant conditioning on thousands of rats and mice, ethology was being developed in Europe. Ethology is the study of animal behaviour in natural environments and the primary concern of the ethologists is instinctive or innate behaviour (Eibl-Eibesfeldt and Kramer, 1958). Essentially, ethologists believe that the secrets to behaviour are found in the animals genes and in the way the genes have been modified during evolution to deal with particular environments. The ethological trend originated with Whitman (1898), who regarded instincts as congenital reactions which are so constant and characteristic for each species that, like morphological structures, they may he of taxonomic significance. A similar opinion was held by Heinroth (1918). He trained newly hatched fledglings in isolation from adult birds of their own species and found that instinctive movements such as preening, shaking, and scratching were performed by young birds without observing other birds.

Understanding the mechanisms and programming that result in innate behavioural patterns and the motivations behind why animals behave the way they do is the primary focus of ethologists. Konrad Lorenz (1939, 1965, 1981) and Niko Tinbergen (1948, 1951) cataloged the behaviour of many animals in their natural environments. Together they developed the ethogram. An ethogram is a complete listing of all the behaviours that an animal performs in its natural environment. The ethogram includes both innate and learned behaviours.

An interesting contribution to ethology came from studies on egg-rolling behaviour in the greylag goose (Lorenz, 1965, 1981). He observed that when a brooding goose notices an egg outside her nest, an innate instinctive program is triggered to retrieve it. The goose fixates on the egg, rises to extend her neck and bill out over it, then gently rolls it back to the nest. This behaviour is performed in a highly mechanical way If the egg is removed as the goose begins to extend her neck, she still completes the pattern of rolling the nonexistent egg back to the nest. Lorenz (1939) and Tinbergen (1948) termed this a 'fixed action pattern." Remarkably, Tinbergen also discovered that brooding geese can be stimulated to perform egg rolling on such items as beer cans and baseballs. The fixed action pattern of rolling the egg back to the nest can be triggered by anything outside the nest that even marginally resembles an egg. Tinbergen realised that geese possess a genetic-releasing mechanism for this fixed action pattern. Lorenz and Tinbergen called the object that triggers the release of a fixed action pattern "sign stimuli." When a mother bird sees the gaping mouth of her young, it triggers the maternal feeding behaviour and the mother feeds her young. The gaping mouth is another example of sign stimuli that acts as a switch and turns on the genetically determined program (Herrick, 1908; Tinbergen, 1951).

Ethologists also explained the innate escape response of newly hatched goslings. When goslings are tested with a cardboard silhouette in the shape of a hawk moving overhead, it triggers a characteristic escape response. The goslings will crouch or run. However, when the silhouette is reversed to look like a goose, there is no effect (Tinbergen, 1951). Several members of the research community doubted the existence of such a hard-wired instinct because other scientists failed to repeat these experiments (Hirsh *et al.*, 1955). Recently Canty and Gould (1995) repeated the classic experiments and explained why the other experiments failed. In the first place, goslings only respond to the silhouette when they are under 7 days old. Second, a large silhouette which casts a shadow must be used; third, goslings respond to the perceived predator differently depending on the circumstances. For example, birds tested alone try to run away from the hawk silhouette and birds reared and tested in groups tend to crouch (Canty and Gould,

1995). Nevertheless, fear is likely to be the basis of the response. Ducklings were shown to have higher heart rate variability when they saw the hawk silhouette (Mueller and Parker, 1980). Research by Balaban (1997) indicates that species-specific vocalisations and head movements in chickens and quail are controlled by distinct cell groups in the brain. To prove this, Balaban transplanted neural tube cells from developing quail embryos into chicken embryos. Chickens hatched from the transplanted eggs exhibited species-specific quail songs and bobbing head movements.

Do similar fixed action patterns occur in mammals? Fentress (1973) conducted an experiment on mice which clearly showed that animals have instinctive species-specific behaviour patterns which do not require learning. Day-old baby mice were anesthetised and had a portion of their front legs amputated. Enough of the leg remained that the mice could easily walk. The operations were performed before the baby mice had fully coordinated movements so there was no opportunity for learning. When the mice became adults, they still performed the species-specific face-washing behaviour; normal mice close their eye just prior to the foreleg passing over the face, and in the amputees the eye still closed before the nonexistent paw hit it. The amputees performed the face- washing routine as if they still had their paws. Fentress (1973) concluded that the experiment proved the existence of instincts in mammals.

The Science of Behaviour Today

Two years after the Brelands article, Jerry Hirsh (1963) at the University of Illinois wrote a paper emphasizing the importance of studying individual differences. He wrote, "Individual differences are no accident. They are generated by properties of organisms as fundamental to behaviour science as thermodynamic properties are to physical science." Today, scientists recognise the contributions of both the Skinnerian and the ethologists approach to understanding behaviour Modern neuroscience supports Darwin's view on behaviour. Bird and mammal brains are constructed with the same basic design. They all have a brain stem, limbic system, cerebellum, and cerebral cortex. The cerebral cortex is the part of the brain used for thinking and flexible problem solving. The major difference between the brains of people and animals is in the size and complexity of the cortex. Primates have a larger and more complex cortex than a dog or a pig; pigs have a more complex cortex than a rat or a mouse. Furthermore, all animals possess innate species-specific motor patterns which interact with experience and learning in the formation of behaviour.

Certain behaviours in both wild and domestic animals are governed largely by innate (hard-wired) programs; however, experiencing and learning are the most important factors in other behaviours.

A basic principle to remember is that animals with large, complex brains are less governed by innate behaviour patterns. For example, bird behaviour is governed more by instinct than that of a dog, whereas an insect would have more hard-wired behaviour patterns than that of a bird. This principle was clear to Yerkes (1905) who wrote:

> *Certain animals are markedly plastic or voluntary in their behaviour, others are as markedly fixed or instinctive. In the primates plasticity has reached its highest known stage of development; in the insects fixity has triumphed, instinctive action is predominant. The ant has apparently sacrificed adaptability to the development of ability to react quickly, accurately and uniformly in a certain way Roughly, animals might he separated into two classes: those which are in high degree capable of immediate adaptation to their conditions, and those that are apparently automatic since they depend upon instinct tendencies to action instead of upon rapid adaptation.*

Interactions between Genetics and Experience

Some behaviour patterns are similar between different species, and some are found only in a particular species. For example, the neural programs that enable animals to walk are similar in most mammals (Melton, 1991). On the other hand, courtship rituals in birds are very species-specific (Nottebohm, 1977). Some innate behaviour patterns are very rigid and experience has little effect on them; other instinctive behaviours can be modified by learning and experience. The flehmann, or lip curl response of a bull when he smells a cow in estrus, and the kneel-down (lordois) posture of a rat in estrus are examples of behaviours that are rigid. Suckling of the mother by newborn mammals is another example of a hard-wired behavioural system. Suckling behaviour does not vary Newborn mammals suckle almost anything put in their mouth.

An example of an innate behaviour that is affected by learning is burrowing behaviour in rats. Boice (1977) found that wild Norway rats and albino laboratory rats both dig elaborate burrows. Learning has some effect on the efficiency of burrowing, but the configuration of the burrows was the same for both the wild and domestic rats. The albino laboratory rats dug excellent burrows the first time they were exposed to an outdoor pen. Nest building in sows is another example of the interaction between instinct and learning. When a sow is having her first litter, she has an uncontrollable urge to build a nest. Nest building is hard-wired and hormonally driven because prostaglandin F_2a injections will induce it in sows (Widowski and Curtis, 1989). However, sows earn from experience how to build a better nest with each successful litter.

Other behaviours are almost entirely learned. Some seagulls learn to drop shellfish on rocks to break them open, while others drop them on the road and let cars break them open (Grandin, 1995). Many animals ranging from apes to birds use tools to obtain food. Griffin (1994) and Dawkins (1993) provide many examples of complex learned behaviours and flexible problem solving in animals.

Innate behaviours used for finding food, such as grazing, scavenging, or hunting, are more dependent on learning than behaviours used to consume food. Mating behaviour, nesting, eating, and prey-killing behaviours tend to be governed more by instinct (Gould, 1977). The greater dependence on learning to find food makes animals in the wild more flexible and able to adapt to a variety of environments. Behaviours used to kill or consume food can be the same in any environment. Mayr (1974) called these different behavioural systems "open" or "closed" to the effects of experience. A lion hunting her prey is an example of an open system. The hunting female lion recognises her prey from a distance and carefully stalks her approach. Herrick (1910) wrote, "the details of the hunt vary every time she hunts; therefore, no combination of simple reflex arcs laid down in the nervous system will be adequate to meet the infinite variations of the requirements for obtaining food:'

Complex Interactions

Some of the interactions between genetics and experience have very complex effects on behaviour. In birds, the chaffinch learns to sing its species-specific song even when reared in a sound-proof box where it is unable to hear other birds (Nottebohm, 1970, 1979). However, when chaffinches are allowed to hear other birds sing, they develop a more complex song. The basic pattern of canary song emerges even in the absence of conspecific (flock-mate) auditory models (Metfessel, 1935; Poulsen, 1959). Young canaries imitate the song of adult canaries they can hear, and when reared in groups they develop song patterns that they all share (Nottebohm, 1977). Many birds, such as the white crowned sparrow chaffinch, and parrot, can develop local song dialects (Nottebohm *et al.*, 1976). Sparrows are able to learn songs by listening to recordings of songs with either pure tones or harmonic overtones. Birds trained with harmonic overtones learned to sing songs with harmonic overtones, but 1 year later, 85% of their songs reverted back to innate pure tone patterns (Nowicki and Marler, 1988). Further experiments by Mundinger (1995) attempted to determine the relative contribution of genetics and learning in bird song. Inbred lines of roller and border canaries were used in this study along with a hybrid cross of the two.

The rollers were cross fostered to border hens and vice versa to control for effects of maternal behaviour. The roller and border males preferred to sing innate song patterns instead of copying their tutors. The hybrids preferred to learn some of both songs. Furthermore, canaries are capable of learning parts of an alien song but have a definite preference for their own songs. Comparing these animals to those in Brelands and Brelands (1961) exhibits, birds can be trained to sing a different song, but genetically determined patterns have a strong tendency to override learning. In reviewing all this literature, it became clear that innate patterns in mammals can be overridden. Unfortunately the animals tend to revert back to innate behaviour patterns.

The Paradox of Novelty

Novelty is anything new or strange in an animal's environment. Novelty is a paradox because it is both fear-provoking and attractive. Paradoxically it is most fear-provoking and attractive to animals with a nervous, excitable temperament. Skinner (1922) wrote that a flighty animal such as the pronghorn antelope will approach a person lying on the ground waving a red flag. Einarsen (1948) further observed that some wild animals will approach various large, dangerous objects such as a power steam shovel. In more recent studies, Kruuk (1972) also observed attraction and reaction to novelty in Thompson's gazelles in Africa. In small groups, Thompson's gazelles are most watchful for predators (Elgar, 1989). Animals that survive in the wild by flight are more attentive to novelty than more placid animals. Gazelles can also distinguish between a dangerous hunting predator and one that is not hunting. The most dangerous predators attract the highest degrees of attraction in the Thompson's gazelle. They often move close to a cheetah when the cheetah is not hunting. Furthermore, when predators walk through a herd of Thompson's gazelles, the size of the flight zone varies depending on the species of predator.

Reaction to Novelty

Confronted with sudden novelty, highly reactive animals are more likely to have a major fear reaction. Examples of sudden novelty include being placed in a new cage, transport in a strange vehicle, an unexpected loud noise, or being placed in an open field. Using various experimental environments, Hennessy and Levine (1978) found that rats show varying degrees of stress and stress hormone levels proportional to the degree of novelty of the environment they are placed in; a glass jar is totally novel in appearance compared to the lab cage to which the animal was accustomed. Being placed in a glass jar was more stressful for rats than a clean lab cage with no bedding.

Livestock and Poultry Reaction to Novelty

Studies of the reaction to novelty in farm animals have been conducted by Moberg and Wood (1982), Stephens and Toner (1975), and Dantzer and Mormede (1983). When calves are placed in an open field test arena that is very dissimilar from their home pen, they show the highest degrees of stress (Dantzer and Mormede, 1983). Calves raised indoors were more stressed by an outdoor arena and calves raised outdoors were more stressed by an indoor arena. The second author is painfully familiar with similar responses in horses. When horses are taken to the mountains for the first time, a well-trained riding horse that is accustomed to many different show rings may panic when it sees a butterfly or hears a twig snapping on a mountain trail.

Genetic Factors and the Need for Novelty

In mammals and birds, normal development of the brain and sense organs requires novelty and varied sensory input. Nobel prize winning research of Hubel and Wiesel (1970) showed that the visual system is permanently damaged if kittens do not receive varied visual input during development. When dogs are raised in barren and nonstimulating environments they are also more excitable (Walsh and Cummins, 1975; Melzak and Burns, 1965). Schultz (1965) stated, "when stimulus variation is restricted central regulation of threshold sensitivities will function to lower sensory thresholds." Krushinski (1960) studied the influence of isolated conditions of rearing on the development of passive defense reactions (fearful aggression) in dogs and found that the expression of a well-marked fear reaction depends on the genotype of the animal. Airedales and German shepherds were reared under conditions of freedom (in homes) and in isolation (in kennels). Krushinski (1960) found that the passive defense reaction developed more acutely and reached a greater degree in the German shepherds kept in isolation compared to the Airedales. In general, animals reared in isolation become more sensitive to sensory stimulation because the nervous system attempts to readjust for the previous lack of stimulation.

In an experiment with chickens, Murphy (1977) found that chicks from a flighty genetic line were more likely to panic when a novel ball was placed in their pen, but they were also more attracted to a novel food than birds from a calm line. Cooper and Zubeck (1958), and Henderson (1968) found that rats bred to be dull greatly improved in maze learning when housed in a cage with many different objects; however, enriched environments had little effect on the rats bred for high intelligence. Greenough and Juraska (1979) found that rearing rats in an environment with many novel objects improves learning and resulted in increased growth of dendrites, which are nerve endings in the brain.

Pigs raised in barren concrete pens also seek stimulation (Grandin, 1989a,b; Wood-Gush and Vestergaard, 1991; Wood-Gush and Beilharz, 1983). Piglets allowed to choose between a familiar object and a novel object prefer the novel object (Wood-Gush and Vestergaard, 1991). Pigs raised on concrete are strongly attracted to novel objects to chew on and manipulate. The first author has observed that nervous, excitable hybrid pigs often chew and bit vigorously on boots or coveralls. This behaviour is less common in placid genetic lines of pigs.

Although hybrid pigs are highly attracted to novelty, tossing a novel object into their pen will initially cause a strong flight response. Compared to calm genetic lines, nervous-hybrid pigs pile up and squeal more when startled. Pork producers report that nervous, fast-growing, lean hybrid pigs also tail-bite other pigs more often than calmer genetic lines Of pigs. jail biting occurs more often when pigs are housed on a concrete slatted floor which provides no opportunity for rooting.

Strong attraction or strong reaction to novelty has also been observed by the first author in cattle. Cattle will approach and lick a piece of paper laying on the ground when they can approach it voluntarily. However, the same piece of paper blowing in the wind may trigger a massive flight response. Practical experience by both authors suggests that highly reactive horses are more likely to engage in vices such as cribbing or stall weaving when housed in stalls or runs where they receive little exercise. Denied variety and novelty in their environments, highly reactive animals adapt poorly compared to animals from calmer genetic lines (Price, 1984).

In summary, in both wild and domestic animals novelty is both highly feared and necessary Novelty is most desirable when animals can approach it slowly. Unfortunately, novelty is also fear-provoking when animals are suddenly confronted with it.

Temperament

In animals as diverse as rats, chickens, cattle, pigs, and humans, genetic factors influence differences in temperament . Some individuals are wary and fearful and others are calm and placid. Boissy (1995) stated, fearfulness is a basic psychological characteristic of the individual that predisposes it to perceive and react in a similar manner to a wide range of potentially frightening events]' In all animals, genetic factors influence reactions to situations which cause fear; therefore, temperament is partially determined by an individual animal's fear response. Rogan and LeDoux (1996) suggest that fear is the product of a neural system that evolved to detect danger and that it causes an animal to make a response to protect itself. Plomin and Daniels (1987) found a substantial genetic influence on shyness (fearfulness) in human children. Shy behaviour in novel situations is considered a stable psychological characteristic of certain individuals. Shyness is also suggested to be among the most heritable dimensions of human temperament throughout the life span.

In an experiment designed to control for maternal effects on temperament and emotionality, Broadhurst (1960) conducted cross-fostering experiments on Maudsley Reactive (MR) and Non Reactive (MNR) rats. These lines of rats are genetically selected for high or low levels of emotional reactivity The results showed that maternal effects were not great enough to completely mask the temperament differences between the two lines (Broadhurst, 1960). Maternal effects can affect temperament, but they are not great enough to completely change the temperament of a cross-fostered animal which has a temperament that is very different from that of the foster mother. In extensive review of the literature, Broadhurst (1975) examined the role of heredity in the formation of behaviour and found that differences in temperament between rats persist when the animals are all raised in the same environment.

Measuring Fear-Based Behaviours

One method of testing fearfulness is the open field test (Hall, 1934). Sudden placement of an animal in an open field test arena is used to measure differences in fearfulness. Open field testing has shown differences in fearfulness between different genetic lines of animals. The test arena floor is usually marked in a grid to measure how much the animals walk around and explore. Huck and Price (1975) showed that domestic rats are less fearful and will walk round the open held more than wild rats. Price and Loomis (1973) explained that some genetic strains of rats are less fearful and explore an open field arena more than others. Eysenck and Broadhurst (1964) found that rodents with high emotional reactivity are more fearful and explore the open field less compared to placid genetic lines.

In their study of genetic effects on behaviour, Fuller and Thompson (1978) found that "simply providing the same defined controlled environment for each genetic group is not enough. Conditions must not only be uniform for all groups, but also favourable to the development of the behaviour of interest." For example, in wartime Russia, Krushinski (1960) investigated the ability of dogs to be trained for the antitank service or as trail dogs trained to track human scent. The dogs were tied to a spike driven into the ground, and the person who regularly looked after them would let them lick from a bowl of food and then summon the dog to follow the man as he retreated 10 to 15 meters.

The dog's activity was measured with a pedometer for the next 2 minutes. The most active dogs were found to be the best antitank dogs. They were also fearless. In the antitank service, dogs were trained to run up to a tank and either run along side of it or penetrate under the caterpillars of the tank. In order to do this, the dogs had to overcome their natural fear of a tank moving toward them at high speed. The less active dogs (as measured by the pedometer) were found to make the best trailer dogs. They slowly followed a trail and kept their noses carefully to the scent while negotiating the corners and turns on the trail. The more active dogs trailed at too high a speed and often jumped the corners and turns in the trail, which sometimes resulted in switching to another trail.

Mahut (1958) demonstrated an example of differences in fear responses between beagles and terriers. When frightened, beagles freeze and terriers run around frantically In domestic livestock, measuring fear reactions during restraint or in an open field test has revealed differences in temperament both between breeds and between individuals within a breed (Grandin, 1993a; Tulloh, 1961; Dantzer and Mormede, 1983; Murphey *et al.*, 1980b, 1981). The fearful, flighty animals become more agitated and struggle more violently when restrained for vaccinations and other procedures (Fordyce *et al.*, 1988; Grandin, 1993a). Fear is likely to be the main cause of agitation during restraint in cattle, horses, pigs, and chickens. Genetic effects on behaviour during transport, handling, and restraint of these animals are further discussed.

Species Differences in Fear Reactions

In an open field test, frightened rodents often stay close to the arena walls, whereas frightened cattle may run around wildly and attempt to escape. Rodents stay close to the walls because they naturally fear open spaces, whereas cattle run around wildly because they fear separation from the herd. This is an example of differences between species in their response to a similar fear- provoking situation. Fear can be manifested in many different ways. For example, a frightened animal may run around frantically and try to escape in one situation, while in another situation the same animal may freeze or limit its movement. Chickens often freeze when handled by humans. Jones (1984) called this "tonic immobility." The chickens become so frightened that they cannot move. Forceful capture of wild animals can sometimes inflict fatal heart damage. Wildlife biologists call this capture myopathy In summary, much is known about the complex phenomenon of fear, but many questions still remain.

Biological Basis of Fear

Genetics influences the intensity of fear reactions. Genetic factors can also greatly reduce or increase fear reaction in domestic animals (Price, 1984; Parsons, 1988; Flint *et al.*, 1995). Research in humans has clearly revealed some of the genetic mechanisms which govern the inheritance of anxiety (Lesch *et al.*, 1996). LeDoux (1992) and Rogan and LeDoux (1996) state that all vertebrates can be fear-conditioned. Davis (1992) recently reviewed studies on the biological basis of fear. Overwhelming evidence points to the amygdala as the fear centre in the brain. A small bilateral structure located in the limbic system, the amygdala is where the triggers for flight or fight" are located. Electrical stimulation of the amygdala is known to increase stress hormones in rats and cats (Matheson *et al.*, 1971; Setckleiv *et al.*, 1961); destroying the amygdala can make a wild rat tame and reduce its emotionality (Kemble *et al.*, 1984). Destroying the amygdala also makes it impossible to provoke a fear response in animals (Davis, 1992). Blanchard and Blanchard (1972) showed that rats lose all of their fear of cats when the amygdala is lesioned. Furthermore, when a rat learns that a signal light means an impending electric shock, a normal response is to freeze. Destroying the amygdala will eliminate this response (Blanchard and Blanchard, 1972; LeDoux *et al.*, 1988, 1990). Finally, electrical stimulation of the amygdala makes humans feel fearful (Gloor *et al.*, 1981). Animal studies also show that stimulation of the amygdala triggers a pattern of responses from the autonomic nervous system similar to that found in humans when they feel fear (Davis, 1992).

Heart rate, blood pressure, and respiration also change in animals when the flight or fight response is activated (Manuck and Schaefer, 1978). All these autonomic functions have neural circuits to the amygdala. Fear can be measured in animals by recording changes in autonomic activity In humans, Manuck and Schaefer (1978) found tremendous differences in cardiovascular reactivity in response to stress, reflecting a stable genetic characteristic of individuals.

Fearfulness and Instinct

Fearfulness and instinct can conflict. This principle was observed firsthand by the second author during his experience raising Queensland Blue Heeler dogs. Annie's first litter was a completely novel experience because she had never observed another dog giving birth or nursing pups. She was clearly frightened when the first pup was born and it was obvious that she did not know what the pup was; however, as soon as she smelled it her maternal instinct took over and a constant uncontrollable licking began. Two years later, Annie's daughter Kay had her first litter. Kay was more fearful than her mother and her highly nervous temperament overrode her innate licking program. When each pup was born Kay ran wildly around the room and would not go near them. The second author had to intervene and place the pups under Kay's nose; otherwise, they may have died. Kay's nervous temperament and fearfulness were a stronger motivation than her motherly instinct.

Nervous System Reactivity Changed by the Environment

Raising young animals in barren environments devoid of variety and sensory stimulation will have an effect on development of the nervous system. It can cause an animal to be more reactive and excitable when it becomes an adult. This is a long-lasting, environmentally induced change in how the nervous system reacts to various stimuli. Effects of deprivation during early development are also relatively permanent. Melzak and Burns (1965) found that puppies raised in barren kennels developed into hyperexcitable adults. In one experiment the deprived dogs reacted with ~diffuse excitement" and ran around a room more than control dogs raised in homes by people. Presenting novel objects to the deprived dogs also resulted in diffuse excitement." Furthermore, the EEGs of the kennel-raised dogs remained abnormal even after they were removed from the kennel (Melzak and Burns, 1965). Simons and Land (1987) showed that the somatosensory cortex in the brains of baby rats will not develop normally if sensory input was eliminated by trimming their whiskers. A lack of sensory input made the brain hypersensitive to stimulation. The effects persisted even after the whiskers had grown back.

Development of emotional reactivity of the nervous system begins during early gestation. Denenberg and Whimbey (1968) showed that handling a pregnant rat can cause her offspring to be more emotional and explore less in an Open field compared to control animals. This experiment is significant because it shows that handling the pregnant mother had the opposite effect on the behaviour of the infant pups. Handling and possibly stressing the pregnant mothers changed the hormonal environment of the fetus which resulted in nervous offspring. However, handling newborn rats by briefly picking them up and setting them in a container reduced emotional reactivity when the rats became adults (Denenberg and Whimbey 1968). The handled rats developed a calmer temperament.

The adrenal glands are known to have an effect on behaviour (Fuller and Thompson, 1978). The inner portions of the adrenals secrete the hormones adrenaline and noradrenaline, while the outer cortex secretes the gender hormones androgens and oestrogens (reproductive hormones), and various corticosteroids (stress hormones). Yeakel and Rhoades (1941) found that Hall's (1938) emotional rats had larger adrenals and thyroids compared to the nonemotional rats. Richter (1952, 1954) found a decrease in the size of the adrenal glands in Norway rats accompanied by domestication. Several line and strain differences have been found since these early reports. Furthermore, Levine (1968) and Levine *et al.* (1967) showed that brief handling of baby rats reduces the response of the adrenal gland to stress. Denenberg *et al.* (1967) concluded that early handling may lead to major changes in the neuroendocrine system.

Changing Reactivity versus Taming

Adult wild rats can be tamed and become accustomed to handling by people (Galef, 1970). This is strictly learned behaviour. Taming full-grown wild animals to become accustomed to handling by people will not diminish their response to a sudden novel stimulus. This principle was demonstrated by Grandin *et al.* (1994) in training wild antelope at the Denver Zoo for low- stress blood testing. Nyala are African antelope with a hair-trigger flight response used to escape from predators. During handling in zoos for veterinary treatments, nyala are often highly stressed and sometimes panic and injure themselves. Over a period of 3 months, Grandin *et al.* (1995) trained nyala to enter a box and stand quietly for blood tests while being fed treats. Each new step in the training had to be done slowly and carefully Ten days were required to habituate the nyala to the sound of the doors on the box being closed.

All the training and petting by zoo keepers did not change the nyala's response to a sudden, novel stimulus. When the nyala saw repairman on the barn roof they suddenly reacted with a powerful fear response and crashed into a fence. They had become accustomed to seeing people standing at the perimeter of the exhibit, but people on the roof was novel and very frightening. Sudden movements such as raising a camera up for a picture also caused the nyala to flee.

Domestic versus Wild

Wild herding species show much stronger fear responses to sudden novelty compared to domestic ruminants such as cattle and sheep. Domestic ruminants have attenuated flight responses due to years of selective breeding (Price, 1984). Wild ruminants will learn to adapt in captivity and associate people with food, but when frightened by some novel stimulus they are more likely to panic and injure themselves (Grandin, 1993b, 1997). This is especially likely if they are prevented from fleeing by a fence or other barrier. Principles for training and handling all herding animals are basically similar. Training procedures used on flighty antelope or placid domestic sheep are the same. The only difference is the amount of time required. Grandin (1989c) demonstrated this by training placid Suffolk sheep to voluntarily enter a tilting restraining device in one afternoon, but it took 3 months to train the nyala.

In summary, experience can affect behaviour in two basic ways: by conventional learning or by changing nervous system reactivity Most importantly, environmental conditions (enriched versus barren) have the greatest effect on the nervous systems of young animals.

Neoteny

Neoteny is the retention of the juvenile features in an adult animal. Genetic factors influence the degree of neoteny in individuals. Neoteny is manifested both behaviourally and physically In the forward to "The Wild Canids" (Fox, 1975), Conrad Lorenz adds a few of his observations on neoteny and the problems of domestication:

The problems of domestication have been an obsession with me for many years. On the one hand I am convinced that man owes the life-long persistence of his constitutive curiosity and explorative playfulness to a partial neoteny which is indubitably a consequence of domestication In a curiously analogous manner does the domestic dog owe its permanent attachment to its master to a behavioural neoteny that prevents it from ever wanting to be a pack leader On the other hand, domestication is apt to cause an equally alarming disintegration of valuable behavioural traits and an equally alarming exaggeration of less desirable ones.

Infantile characteristics in domestic animals are discussed by Price (1984), Lambooij and van Putten (1993), Coppinger and Coppinger (1993), Coppinger and Scheider (1993), and Coppinger *et al.* (1987). The shortened muzzle in dogs and pigs is an example. Domestic animals have been selected for a juvenile head shape, shortened muzzles, and other features (Coppinger and Smith, 1983). Furthermore, retaining juvenile traits makes animals more tractable and easy to handle. The physical changes are also related to changes in behaviour.

Genetic studies point to the wolf as the ancestor of domestic dogs (Isaac, 1970). During domestication, domestic dogs have retained many of the infant wolf behaviours. For example, wolf pups bark and yap a lot but adult wolves rarely bark; domestic dogs bark a lot (Fox, 1975; Scott and Fuller, 1965). Wolves have hard-wired instinctive behaviour patterns that determine dominance or submission in social relationships. In domestic dogs, the ancestral social behaviour patterns of the wolf are fragmented and incomplete. Frank and Frank (1982) observed that the rigid social behaviour of the wolf has disintegrated into "an assortment of independent behavioural fragments." Malamutes raised with wolf pups fail to read the social behaviour signals of the wolf pups. Further comparisons found that the physical development of motor skills is slower in the malamute. Goodwin *et al.* (1997) studied 10 different dog breeds which ranged from German shepherds and Siberian huskies to bulldogs, cocker spaniels, and terriers. They found that the breeds which retained the greatest repertoire of wolf-like social behaviours were the breeds that physically resembled wolves, such as German shepherds and huskies. Barnett *et al.* (1979) and Price (1985) both conclude that experience can also cause an animal to retain juvenile traits. Gould (1977) also considered the effects of neoteny and stated that neoteny is determined by changes in a few genes that determine the timing of different developmental stages.

Overselection for Specific Traits

Countless examples of serious problems caused by continuous selection for a single trait can be found in the medical literature (Steinberg *et al.*, 1994; Dykman *et al.*, 1969). People with experience in animal husbandry know that overselection for single traits can ruin animals. Good dog breeders know this. Sometimes traits that appear to be unrelated are in fact linked. Wright (1922, 1978) demonstrated this clearly by continuous selection for hair colour and hair patterns in inbred strains of guinea pigs. Depressed reproduction resulted in all the strains. Furthermore, differences in temperament, body conformation, and the size and shape of internal organs were found.

Belyaev (1979) further showed that continuous selection for a calm temperament in foxes resulted in negative effects on maternal behaviour and neurological problems. The fox experiments also found graded changes in many traits over several years of continuous selection for tame behaviour. Physiological and behavioural problems increased with each successive generation. In fact, some of the tamest foxes developed abnormal maternal behaviour and cannibalized their pups. Belyaev *et al.* (1981) called this "destabilising selection," in contrast to "stabilising selection" found in nature (Dobzhansky 1970; Gould, 1977).

There are also countless examples in the veterinary medical literature of abnormal bone structure and other physiological defects caused by overselecting for appearance traits in dog breeds (Ott, 1996). The abnormalities range from bulldogs with breathing problems to German shepherds with hip problems. Scott and Fuller (1965) reported the negative effects of continuous selection for a certain head shape in cocker spaniels:

In our experiments we began with what were considered good breeding stocks, with a fair number of champions in their ancestry. When we bred these animals to their close relatives for even one or two generations, we uncovered serious defects in every breed. . .Cocker spaniels are selected for a broad forehead with prominent eyes and a pronounced "stop," or angle, between the nose and forehead. When we examined the brains of some of these animals during autopsy, we found that they showed a mild degree of hydroencephaly; that is, in selecting for skull shape, the breeders accidentally selected for a brain defect in some individuals. Besides all this, in most of our strains only about 50 percent of the females were capable of rearing normal, healthy litters, even under nearly ideal conditions of care.

Overselection in Livestock

Single-minded selection for production traits such as rapid gain and leanness resulted in pigs and cattle with more excitable temperaments (Grandin, 1994). Compared to the older genetic lines with more hack fat, observations by the first author on thousands of pigs indicate that lean hybrids are more excitable and difficult to drive through races. Lean hybrid pigs also have a greater startle response. Separating a single animal from the group is more difficult. Recent research conducted in our laboratory has shown that cattle with an excitable temperament have lower weight gains and more meat quality problems (Voisinet *et al.*, 1997a,b). This research illustrates that selection away from a very excitable temperament would be beneficial. However, overselection for an excessively calm temperament could possibly result in some unknown detrimental trait.

Links between Different Traits

Casual observations by the first author also indicate that the most excitable, flighty pigs and cattle have a long, slender body with fine bones. Some of the lean hybrid pigs have weak legs and a few of the normally brown-eyed pigs now have blue eyes. Blue eyes are often associated with neurological problems (Bergsma and Brown, 1971; Schaible, 1963). Furthermore, pigs and cattle with large, bulging muscles often have a calmer temperament than lean animals with less muscle definition. However, animals with the muscle hypertrophy trait (double muscling) have a more excitable temperament (Holmes *et al.*, 1972). Double muscling is extreme abnormal muscling and it might have the opposite effect on temperament compared to normal muscling.

Another example of apparently unrelated traits being linked is deafness in dogs of the pointer breed selected for nervousness (Kllen *et al.*, 1987, 1988). There appears to be a relationship between thermoregulation and aggressiveness. Wild mice selected for aggressiveness used larger amounts of cotton to build their nests than mice selected for low aggression (Sinyter *et al.*, 1995). This effect occurred in both laboratory and wild Strains of mice.

Researchers using high-tech "knockout" gene procedures have been frustrated by the complexity of genetic interactions. In this procedure, genes are knocked out in a gene-targeting procedure whereby a gene is prevented from performing its normal function. The knockout experiments have shown that blocking different genes can have unexpected effects on behaviour. In one experiment, superaggressive mice were created when genes involved with learning were inactivated (Chen *et al.*, 1994). The mutant mice had little or no fear and fought until they broke their backs. In another experiment the knockout mutants demonstrated normal behaviour until they had pups, and failed to care for them (Brown *et al.*, 1996). In still another experiment, Konig *et al.* (1996) disabled the gene that produces enkephalin (a brain opioid substance) and found unexpected results. Enkephalin is a substance normally involved in pain perception; however, the mice that were deficient in this substance were very nervous and anxious. They ran frantically around their cages in response to noise. The bottom line conclusion from several different knockout experiments is that changing one gene has unexpected effects on other systems. Traits are linked, and it may be impossible to completely isolate single gene effects. Researchers warn that one must be careful not to jump to a conclusion that they have found an '~aggression gene" or a "maternal gene" or an "anxiety gene." To use an engineering analogy, one would not conclude that they had found the "picture center" in a television set after they cut one circuit inside the set that ruined the picture. Gerlai (1996) and Crawley (1996) also warn that knocking out the same gene in two different species may have different effects on behaviour. This is due to the complex interactions between many different genes. Twenty years ago behavioural geneticists concluded that the inheritance of behaviour is complex. Fuller and Thompson (1978) concluded, "It has been found repeatedly that no one genetic mechanism accounts exclusively for a particular kind of behavtor.

Random Factors

behavioural geneticists have discovered that it is impossible to completely control variation in some traits. Gartner (1990) found that breeding genetically similar inbred lines of rats failed to stop weight fluctuations. Even under highly standardised laboratory conditions, body weights continued to fiuctuate between animals. Pig breeders have also observed that commercially bred hybrid lines of pigs do not gain weight at the same rate. Random unknown factors affect variability even in genetically identical animals. Factors in utero may be one cause; the other causes are unknown. Darrel Tatum and his students at Colorado State University found both body conformation and meat quality variation in cattle which were 50% English (Bos taurus) and 50% Brabman (Bos indicus). Some animals had more Brahman characteristics, with larger humps and longer ears than others; the body conformation of many of the animals was not half English and half Brahmatn. The characteristics of the meat varied as well; animals that looked more Brahman had tougher meat. The animals had about 10% variation from the body shape and meat characteristics of Brahman half-bloods.

Gartner (1990) concluded that up to 90% of the cause of random variability cannot be explained by differences in the animals' physical environment. In both mice and cattle, random factors affected body weights. Gartner (1990) believes that the random factors may have their influence either before or shortly after fertilization.

The interactions between environmental and genetic factors are complex. Both an animals' genetic makeup and its environment determine how it will behave. In subsequent chapters in this book the interactions of genetics and environment will be discussed in greater detail. Genetics has profound effects on an animal's behaviour.

There is a complex interaction between genetic and environmental factors which determines how an animal will behave. The animal's temperament is influenced by both genetics and learning. Another principle is that changes in one trait, such as temperament, can have unexpected effects on other apparently unrelated traits. Overselection for a single trait may result in undesirable changes in other behavioural and physical traits.

Sleep (non-human)

Sleep in non-human animals refers to how the behavioural and physiological state of sleep, mainly characterised by reversible unconsciousness, non-responsiveness to external stimuli, and motor passivity, appears in different categories of animals.

***Figure:** A sleeping cat. The upright ears and the body position suggest the cat is experiencing NREM sleep. In REM sleep, the ears would have been tucked in and the skeletal muscles would have been relaxed due to the functional paralysis signifying REM sleep.*

Rats kept from sleeping die within a couple of weeks, but the exact function of sleep is still unknown.

Definition

Sleep can follow a physiological or behavioural definition. In the physiological sense, sleep is a state characterised by reversible unconsciousness, special brainwave patterns, sporadic eye movement, loss of muscle tone (possibly with some exceptions; see below regarding the sleep of birds and of aquatic mammals), and a compensatory increase following deprivation of the state. In the behavioural sense, sleep is characterised by non-responsiveness to external stimuli, the adoption of a typical posture, and the occupation of a sheltered site, all of which is usually repeated on a 24-hour basis. The physiological definition applies well to birds and mammals, but in other animals (whose brain is not as complex), the behavioural definition is more often used. In very simple animals, behavioural definitions of sleep are the only ones possible, and even then the behavioural repertoire of the animal may not be extensive enough to allow distinction between sleep and wakefulness.

Sleep in Different Species

Sleep in Invertebrates: Sleep as a phenomenon appears to have very old evolutionary roots. The nematode *C. elegans* is the most primitive organism in which sleep-like states have been observed. Here, a *lethargus* phase occurs in short periods preceding each moult, a fact which may indicate that sleep primitively is connected to developmental processes. Raizen *et al.*'s results furthermore suggest that sleep is necessary for changes in the neural system.

The electrophysiological study of sleep in small invertebrates is complicated. However, even such simple animals as fruit flies appear to sleep, and systematic disturbance of that state leads to cognitive disabilities. There are several methods of measuring cognitive functions in fruit flies. A common method is to let the flies choose whether they want to fly through a tunnel that leads to a light source, or through a dark tunnel. Normally, flies are attracted to light.

But if sugar is placed in the end of the dark tunnel, and something the flies dislike is placed in the end of the light tunnel, the flies will eventually learn to fly towards darkness rather than light. Flies deprived of sleep require a longer time to learn this and also forget it more quickly. If an arthropod is experimentally kept awake longer than it is used to, then its coming rest period will be prolonged. In cockroaches that rest period is characterised by the antennae being folded down and by a decreased sensitivity to external stimuli. Sleep has been described in crayfish, too, characterised by passivity and increased thresholds for sensory stimuli as well as changes in the EEG pattern, markedly differing from the patterns found in crayfish when they are awake.

Sleep in Fishes and Reptiles: Sleep in fishes is not extensively studied. Some species that always live in shoals or that swim continuously (because of a need for ram ventilation of the gills, for example) are suspected never to sleep. There is also doubt about certain blind species that live in caves. Other fishes seem to sleep, however. For example, zebrafish, tilapia, tench, brown bullhead, and swell shark become motionless and unresponsive at night (or by day, in the case of the swell shark); Spanish hogfish and blue-headed wrasse can even be lifted by hand all the way to the surface without evoking a response. A 1961 observational study of approximately 200 species in European public aquaria reported many cases of apparent sleep. On the other hand, sleep patterns are easily disrupted and may even disappear during periods of migration, spawning, and parental care.

***Figure:** A Komodo dragon sleeping.*

Reptiles have been subjected to electrophysiological studies of sleep. That is to say that electrical activity in the brain has been registered when the animals have been asleep.

However, the EEG pattern in reptilian sleep differs from what is seen in mammals and other animals. In reptiles, sleep time increases following sleep deprivation, and stronger stimuli are needed to awaken the animals when they have been deprived of sleep as compared to when they have slept normally. This suggests that the sleep which follows deprivation is compensatorily deeper.

Sleep in Birds: There are significant similarities between sleep in birds and sleep in mammals, which is one of the reasons for the idea that sleep in higher animals with its division into REM and NREM sleep has evolved together with warm-bloodedness. Birds compensate for sleep loss in a manner similar to mammals, by deeper or more intense SWS (slow-wave sleep).

***Figure:** A sleeping Cockatiel*

Birds have both REM and NREM sleep, and the EEG patterns of both have similarities to those of mammals. Different birds sleep different amounts, but the associations seen in mammals between sleep and variables such as body mass, brain mass, relative brain mass, basal metabolism and other factors (see below) are not found in birds. The only clear explanatory factor for the variations in sleep amounts for birds of different species is that birds who sleep in environments where they are exposed to predators have less deep sleep than birds sleeping in more protected environments.

Figure: *A flamingo with at least one cerebral hemisphere awake*

A peculiarity that birds share with aquatic mammals, and possibly also with certain species of lizards (opinions differ about that last point), is the ability for unihemispheric sleep. That is the ability to sleep with one cerebral hemisphere at a time, while the other hemisphere is awake.

When only one hemisphere is sleeping, only the contralateral eye will be shut; that is, when the right hemisphere is asleep the left eye will be shut, and vice versa. The distribution of sleep between the two hemispheres and the amount of unihemispheric sleep are determined both by which part of the brain has been the most active during the previous period of wake—that part will sleep the deepest—and it is also determined by the risk of attacks from predators. Ducks near the perimeter of the flock are likely to be the ones that first will detect predator attacks. These ducks have significantly more unihemispheric sleep than those who sleep in the middle of the flock, and they react to threatening stimuli seen by the open eye.

Opinions partly differ about sleep in migratory birds. The controversy is mainly about whether they can sleep while flying or not. Theoretically, certain types of sleep could be possible while flying, but technical difficulties preclude the recording of brain activity in birds while they are flying.

Sleep in Mammals:

Figure: *Sleeping polar bears*

Sleep Duration: Different animals sleep different amounts. Some animals, such as bats, sleep 18–20 hours per day, while others, including giraffes, sleep only 3–4 hours per day. There can be big differences even between closely related species. There can also be differences between laboratory and field studies: for example, researchers in 1983 reported that captive sloths slept nearly 16 hours a day, but in 2008, when miniature neurophysiological recorders were developed that could be affixed to wild animals, sloths in nature were found to sleep only 9.6 hours a day.

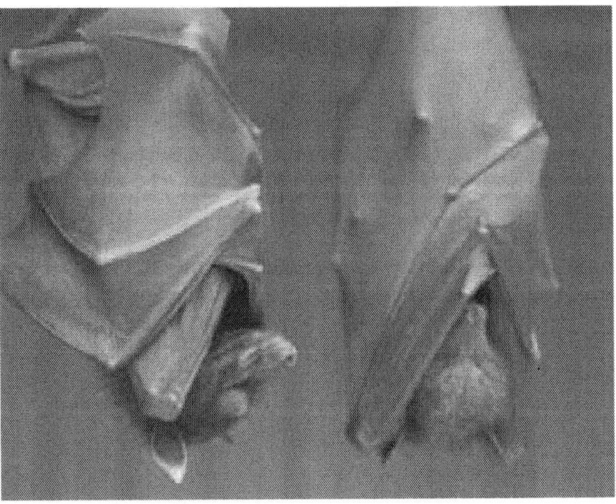

Figure: *Flying foxes, asleep*

As for birds, the main rule for mammals (with certain exceptions, see below) is that they have two essentially different stages of sleep: REM and NREM sleep. An animal's feeding habits are associated with its sleep length. The daily need for sleep is highest in carnivores, lower in omnivores and lowest in herbivores. Humans do not sleep unusually much or unusually little compared to other animals, but we sleep less than many other omnivores.

Many herbivores, like Ruminantia (such as cattle), spend much of their wake time in a state of drowsiness, which perhaps could partly explain their relatively low need for sleep. In herbivores, a direct correlation is apparent between body mass and sleep length; big animals sleep more than smaller ones. This correlation is thought to explain about 25% of the difference in sleep amount between different animals. Also, the length of a particular sleep cycle is associated with the size of the animal; on average, bigger animals will have sleep cycles of longer durations than smaller animals. Sleep amount is also coupled to factors like basal metabolism, brain mass and relative brain mass. Mammals born with well-developed regulatory systems, such as the horse and giraffe, tend to have less REM sleep than the species which are less developed at birth, such as cats and rats. This appears to echo the greater need for REM sleep among newborns than among adults in most mammal species.

Comparative average sleep periods for various mammals in captivity over 24 hours have been given as: horses - 2.9 hours; elephants - 3 plus; cows - 4.0; giraffes - 4.5 hours; humans - 8.0; rabbits - 8.4; chimpanzees - 9.7; Red foxes - 9.8; dogs - 10.1; House mice - 12.5; cats - 12.5; lions - 13.5; platypuses - 14; chipmunks - 15; Giant armadillos - 18.1; and Little brown bats - 19.9 hours. Reasons given for the wide variations include the fact that mammals "that nap in hiding, like bats or rodents tend to have longer, deeper snoozes than those on constant alert." Lions, which have little fear of predators also have relatively long sleep periods, while elephants have to eat most of the time to support their huge bodies. Little brown bats conserve their energy except for the few hours each night when their insect prey are available and platypuses eat a high energy crustacean diet and, therefore, probably don't need to spend as much time awake as many other mammals.

Sleep in Monotremes: Since monotremes, egg-laying mammals, are considered to represent one of the evolutionarily oldest groups of mammals, they have been subject to special interest in the study of mammalian sleep. As early studies of these animals could not find clear evidence for REM sleep, it was initially assumed that such sleep did not exist in monotremes but developed after the monotremes left the rest of the mammals and became a separate, distinct group. However, EEG registrations of the brain stem in monotremes show a firing pattern that is quite similar to the patterns seen in REM sleep in higher mammals. In fact, the largest amount of REM sleep known in any animal is found in the platypus. The average sleep time in a 24-hour period of a platypus is said to be as long as 14 hours. "Why? Maybe because just a little crustacean meal packs a huge caloric punch."

Sleep in Aquatic Mammals: Among others, seals and whales belong to the aquatic mammals. Seals are grouped in earless seals and eared seals, which have solved the problem of sleeping in water differently. Eared seals, like whales, show unihemispheric sleep. The sleeping half of the brain does not awaken when they surface to breathe. When one half of a seal's brain shows slow-wave sleep, the flippers and whiskers on its opposite side are immobile. While in the water, these seals have almost no REM sleep and may go a week or two without it. As soon as they move onto land they switch to bilateral REM sleep and NREM sleep comparable to land mammals, surprising researchers with their lack of "recovery sleep" after missing so much REM.

Figure: *Northern Sea Lion pup with adult female and male, the largest of the eared seals. Habitat: the northern Pacific*

Earless seals sleep bihemispherically like most mammals, under water, hanging at the water surface or on land. They hold their breath while sleeping under water, and wake up regularly to surface and breathe. They can also hang with their nostrils above water and in that position have REM sleep, but they do not have REM sleep underwater.

REM sleep has been observed in the pilot whale, a species of dolphin. Whales do not seem to have REM sleep, nor do they seem to have any problems because of this. One reason REM sleep might be difficult in marine settings is the fact that REM sleep causes muscular atony; that is to say, a functional paralysis of skeletal muscles that can be difficult to combine with the need to breathe regularly.

Chapter 2

The History of Animal Evolution

For many people animals are perhaps the most familiar, and most interesting, of living things. This may be because we are animals ourselves. As such, we have a number of features in common with all the organisms placed in the animal kingdom, and these common features indicate that we have a shared evolutionary history.

All animals and plants are classified as multicellular eukaryotes: their bodies are made up of large numbers of cells, and microscopic inspection of these cells reveals that they contain a nucleus and a number of other organelles. Compared to prokaryotic organisms such as bacteria, plants and animals have a relatively recent evolutionary origin. DNA evidence suggests that the first eukaryotes evolved from prokaryotes, between 2500 and 1000 million years ago. That is, eukaryotes as a taxon date from the Proterozoic Era, the final Era of the Precambrian. Fossils of both simple unicellular and more complex multicellular organisms are found in abundance in rocks from this period of time. In fact, the name "Proterozoic" means "early life".

Plants and animals both owe their origins to endosymbiosis, a process where one cell ingests another, but for some reason then fails to digest it. The evidence for this lies in the way their cells function. Both plant and animal rely on structures called mitochondria to release energy in their cells, using aerobic respiration to produce the energy-carrying molecule ATP. There is considerable evidence that mitochondria evolved from free-living aerobic bacteria: they are the size of bacterial cells; they divide independently of the cell by binary fission; they have their own genome in the form of a single circular DNA molecule; their ribosomes are more similar to those of bacteria than to the ribosomes found in the eukaryote cell's cytoplasm; and like chloroplasts they are enclosed by a double membrane as would be expected if they derived from bacterial cells engulfed by another cell.

Like the plants, animals evolved in the sea. And that is where they remained for at least 600 million years. This is because, in the absence of a protective ozone layer, the land was bathed in lethal levels of UV radiation. Once photosynthesis had raised atmospheric oxygen levels high enough, the ozone layer formed, meaning that it was then possible for living things to venture onto the land.

The oldest fossil evidence of multicellular animals, or metazoans, is burrows that appear to have been made by smooth, wormlike organisms. Such trace fossils have been found in rocks from China, Canada, and India, but they tell us little about the animals that made them apart from their basic shape.

- The Ediacaran animals
- The Cambrian "explosion" and the Burgess Shale
- What caused the Cambrian "explosion"?
- A foot on the land
- The earliest vertebrates
- Appearance of the fish
- The jawless fish
- Colonisation of the land
- Problems encountered in the move to land
- The evolution of amphibians
- What drove amphibian evolution?
- Early reptiles and the amniotic egg
- The early mammals
- Developments in the dinosaur lineage
- Further developments in the early mammals
- Taking wing: Archaeopteryx and the origin of the birds
- The end of the dinosaur age
- The appearance of modern mammal groups.

The Ediacaran Animals

Between 620 and 550 million years ago (during the Vendian Period) relatively large, complex, soft-bodied multicellular animals appear in the fossil record for the first time. While found in several localities around the world, this particular group of animals is generally known as the Ediacaran fauna, after the site in Australia where they were first discovered. The Ediacaran animals are puzzling in that there is little or no evidence of any skeletal hard parts i.e. they were soft-bodied organisms, and while some of them may have belonged to groups that survive today others don't seem to bear any relationship to animals we know. Although many of the Ediacaran organisms have been compared to modern-day jellyfish or worms, they have also been described as resembling a mattress, with tough outer walls around fluid-filled internal cavities - rather like a sponge.

As a group, Ediacaran animals had a flat, quilted appearance and many showed radial symmetry. They ranged in size form 1cm to >1m, and have been classified into three main groups on the basis of their shape: discoidal, frond-like, or ovate-elongate. The large variety of Ediacaran animals is significant, as it suggests there must have been a lengthy period of evolution prior to their first appearance in the fossil record.

The Cambrian "Explosion" and the Burgess Shale

The Ediacaran animals disappear from the fossil record at the end of the Vendian (544 million years ago). In their place we find representatives of almost all the modern phyla recognised today: sponges, jellyfish and corals, flatworms, molluscs, annelid worms, insects, echinoderms and chordates, plus many "lesser" phyla such as nemertean worms. These "modern" organisms appear relatively quickly in the geological time scale, and their abrupt appearance is often described as the "Cambrian explosion" however, bear in mind that the fossil record of the "explosion" is spread over about 30 million years. I keep taking things out of brackets because it is interesting relevant and memorable

One of the most famous assemblages of Cambrian fossils comes from the Burgess Shale of British Colombia. The rocks of the Burgess Shale were laid down in the middle Cambrian, when the "explosion" had already been underway for several million years. They contain familiar animals such as trilobites, molluscs and echinoderms, but also the first appearance of brachiopods, and some odd animals, e.g. *Opabinia*, that may have belonged to extinct phyla. Even an early chordate, *Pikaia*, has been found in this fossil assemblage.

The Burgess Shale fossils are important, not only for their evidence of early variety among animal forms, but also because both soft parts of animals and their hard bodies (i.e. the whole animal) is preserved, and animals that were entirely soft-bodied. Preservation of soft-bodied organisms is rare, and in this case seems to have occurred when the animals were rapidly buried in a mudslide down into deep, anaerobic waters, where there was little bacterial decay. Prior to the discovery of this fossil assemblage, early in the 20th century, there was no evidence of soft-bodied animals from the Cambrian (remember that this is before the Ediacaran fauna were found).

These fossils also provide good evidence of predatory animals (e.g. Anomalocaris), and therefore of complex predator-prey relationships. They also give insights into how evolution might have progressed relatively early in the history of multicellular animals, and in fact some authors view the Cambrian as a period of extreme "experimentation" and diversity.

What Caused the Cambrian "Explosion"?

The cause of the proliferation of animal forms in the Cambrian is a matter of considerable debate among scientists. Some point to the increase in atmospheric oxygen levels that began around 2000 million years ago, supporting a higher metabolic rate and allowing the evolution of larger organisms and more complex body structures. Changed ocean chemistry would have played a part here, allowing for the first time the development of hard body parts such as teeth and supporting skeletons based on calcium carbonate ($CaCO_3$), and also supporting higher levels of primary production as a result of increased concentrations of phosphates and nitrates. The mass extinction that marked the end of the Vendian period would have opened up ecological niches that the new animals exploited, as would habitat changes wrought by continental drift.

Genetic factors were also crucial. Recent research suggests that the period prior to the Cambrian explosion saw the gradual evolution of a "genetic tool kit" of genes (the homeobox or "hox" genes) that govern developmental processes. Once assembled, this genetic tool kit enabled an unprecedented period of evolutionary experimentation — and competition. Many forms seen in the fossil record of the Cambrian disappeared without trace. Future evolutionary change was then limited to acting on the body plans that remained in existence.

Recently many scientists have begun to question whether the Cambrian explosion was a real event, or a reflection of the patchiness of this ancient fossil record. Genetic data suggest that multicellular animals evolved around 1000 million years ago; this is supported by fossil embryos from rocks in China that date back 600 million years. These embryos are more complex than those of simple organisms such as sponges and jellyfish, which suggests that multicellular animals must have evolved much further back in time. In addition, trilobites were a very diverse group even early in the Cambrian, and some scientists suggest that this indicates that the arthropod group must have had a much earlier evolutionary origin.

A Foot on the Land

Whatever their origins, animals may have ventured onto land early in the Cambrian. Previously scientists believed that animals did not begin to colonise the land until the Silurian (440 - 410 million years ago). However, the 2002 discovery of the footprints of animals that scuttled about on sand dunes about 530 million years ago has changed this view. These animals were arthropods, and resembled centipedes about the size of crayfish. They probably didn't live on land, instead coming ashore to mate or evade predators. At this time the only land plants appear to have resembled mosses.

The Earliest Vertebrates

Animals continued to diversify in the Ordovician seas (505 - 440 million years ago). They were mostly invertebrates, including graptolites, which were stick-like branching colonies of tiny animals, together with brachiopods, trilobites, cephalopods, corals, crinoids and conodonts. We now place the conodonts with the chordates, but for a long time they were known only by their tiny, but very common, teeth.

In terms of number of species invertebrates were by far the most common Ordovician animals - as they still are today. However, members of another taxon were also evolving in the Ordovician seas. These were the fish.

Appearance of the Fish

Like the conodonts, fish are members of the chordate phylum because they display certain defining characteristics: a dorsal stiffening rod called the notochord, a dorsal nerve cord, pharyngeal gill slits and a tail that extends beyond the anus. However, fish are placed in the subphylum Vertebrata, because they also show the development of skeletal features such as a backbone, skull, and limb bones.

Not all the modern groups of fish were represented in the Ordovician oceans. At this time only the jawless fish had evolved from a chordate ancestor. The sharks and their relatives and two extinct groups, the placoderms (which had bony plates covering their heads) and the acanthodians (the first known jawed vertebrates, with a skeleton of cartilage) made their appearance in the Silurian. However, neither the sharks nor the agnathans became common until the Devonian. The other two living lineages, the ray-finned (e.g. carp and kahawai) and the lobe-finned fish (e.g. lungfish and the coelacanth), evolved during the Devonian period.

The Jawless Fish

Agnathans, or jawless fish, were the earliest fish: an excellent fossil of *Haikouichthys ercaicunensis* dates back about 530 million years, to the Cambrian. Previously the earliest-known agnathans were dated to around 480 million years ago. Agnathans have traditionally been placed with the vertebrates due to the presence of a skull, although the modern forms such as hagfish lack a vertebral column. The earliest agnathans were Ostracoderms. They were bottom-feeders and were almost entirely covered in armour plates. When the sharks and bony fish began to evolve, around 450 million years ago, most ostracoderms became extinct. Only the lineage that produced the modern hagfish and lampreys survived.

Colonisation of the Land

Fish continued to evolve during the Silurian period (440 - 410 million years ago). At the same time some groups of plants and animals took a major step as they colonised the land for the first time. We are not sure why this advance occurred, but it was probably the result of competition in the marine ecosystems, plus the opportunity to escape predators and the availability of new terrestrial niches.

Arthropods, which had ventured temporarily onto land 100 million years earlier, were the first animals to become more permanent colonists. Fossil footprints made in the sandy flats surrounding temporary lakes dating back about 420 million years have been found in Western Australia.

The arthropods were pre-adapted to life on land. By the time they moved ashore, they had already evolved lighter bodies and slim, strong legs that could support them against the pull of gravity. Their hard outer exoskeletons provided protection and would help to retain water, although the development of a waxy, waterproof cuticle was necessary for efficient water conservation.

Spiders, centipedes and mites were among the earliest land animals. Some of them were giants: the largest was *Slimonia*, the size of a man and a relative of the scorpions. This animal was still too big and too heavy and the walking legs too small to venture onto land for any length of time and so they lived in marginal marine (deltaic) environments.

Problems Encountered in the Move to Land

These early land animals had to solve the same problems that plants faced when they moved to the land: water conservation, gas exchange, reproduction and dispersal, and the fact that water no longer buoyed them up against the pull of gravity. Like plants, animals evolved waterproof external layers, internal gas exchange systems, ways of reproducing that did not involve water, and strong support systems (endoskeletons and exoskeletons) that allowed them to move about on land. Remember that not all animal taxa were equally successful in solving these problems.

The Evolution of Amphibians

By the Devonian period two major animal groups dominated the land: the tetrapods (4-legged terrestrial vertebrates) and the arthropods, including arachnids and wingless insects. The first tetrapods were amphibians, such as *Ichthyostega*, and were closely related to a group of fish known as lobe-finned fish e.g. *Eusthenopteron*. Once thought to be extinct, the coelacanth is a living representative of this group.

Eusthenopteron had a number of exaptations that pre-adapted it to life on land: it had limbs (with digits) that allowed it to move around on the bottom of pools, lungs - which meant it could gulp air at the surface, and the beginnings of a neck. This last is important as a terrestrial predator cannot rely on water current to bring food into its mouth, but must move its head to catch prey. And the bones in *Eusthenopteron*'s fins are almost identical to those in the limbs of the earliest amphibians, an example of homology.

Ichthyostega's skull was almost identical to that of the lobe-finned fish *Eusthenopteron*, a definite neck separated its body from its head, and it retained a deep tail with fins. While *Ichthyostega* had four strong limbs, the form of its hind legs suggests that it did not spend all its time on land.

All modern tetrapods have a maximum of 5 digits on each limb, and are thus said to have a pentadactyl limb. For a long time scientists believed that pentadactyly was the ancestral state for tetrapods. However, careful examination of the fossils of early amphibians such as *Ichthyostega* and *Acanthostega* has revealed the presence of up to 8 toes on each foot!

In addition, these early amphibians were large-bodied animals with strong bodies and prominent ribs - quite different in appearance from modern representatives such as frogs and axolotls.

What Drove Amphibian Evolution?

It was originally believed that the tetrapods evolved during periods of drought, when the ability to move between pools would be an advantage. The animals would also have been able to take advantage of terrestrial prey, such as arthropods. Juvenile animals could avoid predation by the land-based adults by living in shallow water.

However, fossil and geological evidence tells us that the early tetrapods lived in lagoons in tropical regions, so that drought was not an issue. They were unlikely to be feeding on land: arthropods are small and fast-moving, unlikely prey for large, sluggish amphibians. But amphibians that laid their eggs on land, rather than in water, would be at a selective advantage, avoiding predation by aquatic vertebrates (such as other amphibians and fish) on gametes, eggs and hatchlings.

Even today some amphibians e.g. the Eleutherodactylid frogs of Australia and Indonesia lay their eggs in soil on the land. However, they must still be in a moist environment, and the size of the egg is restricted to less than 1.5cm in diameter. This is because the egg is dependent on diffusion alone for gas exchange, and means that the embryo must develop rapidly into a food-seeking larval form rather than undergo prolonged development within the egg.

In the Devonian seas, brachiopods had become a dominant invertebrate group, while the fish continued to evolve, with sharks becoming the dominant marine vertebrates. The placoderms and acanthodian fish were quite diverse during the Devonian, but their numbers then dwindled rapidly and both groups became extinct by the end of the Carboniferous period. Lobe-finned fish also peaked in numbers during the Devonian.

Early Reptiles and the Amniotic Egg

One of the greatest evolutionary innovations of the Carboniferous period (360 - 268 million years ago) was the amniotic egg, which allowed early reptiles to move away from waterside habitats and colonise dry regions. The amniotic egg allowed the ancestors of birds, mammals, and reptiles to reproduce on land by preventing the embryo inside from drying out, so eggs could be laid away from the water. It also meant that in contrast to the amphibians the reptiles could produce fewer eggs at any one time, because there was less risk of predation on the eggs. Reptiles don't go through a larval food-seeking stage, but undergo direct development into a miniature adult form while in the egg, and fertilization is internal.

The earliest date for development of the amniotic egg is about 320 million years ago. However, reptiles didn't undergo any major adaptive radiation for another 20 million years. Current thinking is that these early amniotes were still spending time in the water and came ashore mainly to lay their eggs, rather than to feed. It wasn't until the evolution of herbivory that new reptile groups appeared, able to take advantage of the abundant plant life of the Carboniferous.

Early reptiles belonged to a group called the cotylosaurs. *Hylonomus* and *Paleothyris* were two members of this group. They were small, lizard-sized animals with amphibian-like skulls, shoulders, pelvis and limbs, and intermediate teeth and vertebrae. The rest of the skeleton was reptilian. Many of these new "reptilian" features are also seen in little, modern, amphibians (which may also have direct-developing eggs laid on land e.g. New Zealand's leiopelmid frogs, so perhaps these features were simply associated with the small body size of the first reptiles.

The Early Mammals

A major transition in the evolution of life occurred when mammals evolved from one lineage of reptiles. This transition began during the Permian (286 - 248 million years ago), when the reptile group that included *Dimetrodon* gave rise to the "beast-faced" therapsids. (The other major branching, the "lizard-faced" sauropsids, gave rise to birds and modern reptiles). These mammal-like reptiles in turn gave rise to the cynodonts e.g. *Thrinaxodon* during the Triassic period.

The development of a key mammalian trait, the presence of only a single bone in the lower jaw (compared to several in reptiles) can be traced in the fossil history of this group. It includes the excellent transitional fossils, *Diarthrognathus and Morganucodon*, whose lower jaws have both reptilian and mammalian articulations with the upper. Other novel features found in this lineage include the development of different kinds of teeth (a feature known as heterodonty), the beginnings of a secondary palate, and enlargement of the dentary bone in the lower jaw. Legs are held directly underneath the body, an evolutionary advance that occurred independently in the ancestors of the dinosaurs.

The end of the Permian was marked by perhaps the greatest mass extinction ever to occur. Some estimates suggest that up to 90% of the species then living became extinct. (Recent research has suggested that this event, like the better-known end-Cretaceous event, was caused by the impact of an asteroid.) During the subsequent Triassic period (248 - 213 million years ago), the survivors of that event radiated into the large number of now-vacant ecological niches.

However, at the end of the Permian it was the dinosaurs, not the mammal-like reptiles, which took advantage of the newly available terrestrial niches to diversify into the dominant land vertebrates. In the sea, the ray-finned fish began the major adaptive radiation that would see them become the most species-rich of all vertebrate classes.

Developments in the Dinosaur Lineage

One major change, in the group of reptiles that gave rise to the dinosaurs, was in the animals' posture. This changed from the usual "sprawling" mode, where the limbs jut sideways, to an erect posture, with the limbs held directly under the body. This had major implications for locomotion, as it allowed much more energy-efficient movement.

The dinosaurs, or "terrible lizards", fall into two major groups on the basis of their hip structure: the saurischians (or "lizard-hipped" dinosaurs) and the ornithischians (misleadingly known as the "bird-hipped" dinosaurs). Ornithischians include *Triceratops*, *Iguanodon*, *Hadrosaurus*, and *Stegosaurus*). Saurischians are further subdivided into theropods (such as *Coelophysis* and *Tyrannosaurus* rex) and sauropods (e.g. *Apatosaurus*). Most scientists agree that birds evolved from theropod dinosaurs.

Although the dinosaurs and their immediate ancestors dominated the world's terrestrial ecosystems during the Triassic, mammals continued to evolve during this time.

Further Developments in the Early Mammals

Mammals are advanced synapsids. Synapsida is one of two great branches of the amniote family tree. Amniotes are the group of animals that produce an amniotic egg i.e. the reptiles, birds, and mammals. The other major amniote group, the Diapsida, includes the birds and all living and extinct reptiles other than the turtles and tortoises. Turtles and tortoises belong in a third group of amniotes, the Anapsida. Members of these groups are classified on the basis of the number of openings in the temporal region of the skull.

Synapsids are characterised by having a pair of extra openings in the skull behind the eyes. This opening gave the synapsids (and similarly the diapsids, which have two pairs of openings) stronger jaw muscles and better biting ability than earlier animals. (The jaw muscles of a synapsid are anchored to the edges of the skull opening). Pelycosaurs (like *Dimetrodon* and *Edaphosaurus*) were early synapsids; they were mammal-like reptiles. Later synapsids include the therapsids and the cynodonts, which lived during the Triassic.

Cynodonts possessed many mammalian features, including the reduction or complete absence of lumbar ribs implying the presence of a diaphragm; well-developed canine teeth, the development of a bony secondary palate so that air and food had separate passages to the back of the throat; increased size of the dentary - the main bone in the lower jaw; and holes for nerves and blood vessels in the lower jaw, suggesting the presence of whiskers.

By 125 million years ago the mammals had already become a diverse group of organisms. Some of them would have resembled today's monotremes (e.g. platypus and echidna), but early marsupials (a group that includes modern kangaroos and possums) were also present. Until recently it was thought that placental mammals (the group to which most living mammals belong) had a much later evolutionary origin. However, recent fossil finds and DNA evidence suggest that the placental mammals are much older, perhaps evolving more than 105 million years ago. Note that the marsupial and placental mammals provide some excellent examples of convergent evolution, where organisms that are not particularly closely related have evolved similar body forms in response to similar environmental pressures.

However, despite the fact that the mammals had what many people regard as "advanced" features, they were still only minor players on the world stage. As the world entered the Jurassic period (213 - 145 million years ago), the dominant animals on land, in the sea, and in the air, were the reptiles. Dinosaurs, more numerous and more extraordinary than those of the Triassic, were the chief land animals; crocodiles, ichthyosaurs, and plesiosaurs ruled the sea, while the air was inhabited by the pterosaurs.

Taking Wing: Archaeopteryx and the Origins of the Birds

In 1861 an intriguing fossil was found in the Jurassic Solnhofen Limestone of southern Germany, a source of rare but exceptionally well-preserved fossils. Given the name *Archeopteryx lithographica* the fossil appeared to combine features of both birds and reptiles: a reptilian skeleton, accompanied by the clear impression of feathers. This made the find highly significant as it had the potential to support the Darwinians in the debate that was raging following the 1859 publication of "On the origin of species".

While it was originally described as simply a feathered reptile, *Archaeopteryx* has long been regarded as a transitional form between birds and reptiles, making it one of the most important fossils ever discovered. Until relatively recently it was also the earliest known bird. Lately, scientists have realised that *Archaeopteryx* bears even more resemblance to the Maniraptora, a group of dinosaurs that includes the infamous velociraptors of "Jurassic Park", than to modern birds. Thus the *Archaeopteryx* provides a strong phylogenetic link between the two groups. Fossil birds have been discovered in China that are even older than Archaeopteryx, and other discoveries of feathered dinosaurs support the theory that theropods evolved feathers for insulation and thermo-regulation before birds used them for flight. This is an example of an exaptation.

Closer examination of the early history of birds provides a good example of the concept that evolution is neither linear nor progressive. The bird lineage is messy, with a variety of "experimental" forms appearing. Not all achieved powered flight, and some looked quite unlike modern birds e.g. *Microraptor gui*, which appears to have been a gliding animal and had asymmetric flight feathers on all four limbs, while its skeleton is essentially that of a small dromaeosaur. *Archaeopteryx* itself did not belong to the lineage from which modern birds (*Neornithes*) have evolved, but was a member of the now-extinct *Enantiornithes*. A reconstruction of the avian family tree would show a many-branched bush, not a single straight trunk.

The End of the Dinosaur Age

Dinosaurs spread throughout the world - including New Zealand, which had its own dinosaur fauna - during the Jurassic, but during the subsequent Cretaceous period (145 - 65 million years ago) they were declining in species diversity. In fact, many of the typically Mesozoic organisms - such as ammonites, belemnites, gymnosperms, ichthyosaurs, plesiosaurs, and pterosaurs - were in decline at this time, despite the fact that they were still giving rise to new species.

The origin of flowering plants (the angiosperms) during the early Cretaceous triggered a major adaptive radiation among the insects: new groups, such as butterflies, moths, ants and bees arose and flourished. These insects drank the nectar from the flowers and acted as pollinating agents in the process.

The mass extinction at the end of the Cretaceous period, 65 million years ago, wiped out the dinosaurs along with every other land animal that weighed much more than 25 kg. This cleared the way for the expansion of the mammals on land. In the sea at this time, the fish again became the dominant vertebrate taxon.

The Appearance of Modern Mammal Groups

At the beginning of the Palaeocene epoch (65 - 55.5 million years ago) the world was without larger-sized terrestrial animals. This unique situation was the starting point for the great evolutionary diversification of the mammals, which up until then had been nocturnal animals the size of small rodents. By the end of the epoch, mammals occupied many of the vacant ecological niches. While mammal fossils from this period of time are scarce, and often consist largely of their characteristic teeth, we know that small, rodent-like insectivorous mammals roamed the forests, the first large herbivorous mammals were browsing on the abundant vegetation, and carnivorous mammals were stalking their prey.

The oldest confirmed primate fossils date to about 60 million years ago, in the mid-Palaeocene. The early primates evolved from archaic nocturnal insectivores, something like shrews, and resembled lemurs or tarsiers (the prosimians). They were probably arboreal, living in tropical or subtropical forests. Many of their characteristic features are well suited for this habitat: hands specialised for grasping, rotating shoulder joints, and stereoscopic vision. They also have a relatively large brain size and nails on their digits, instead of claws.

The earliest known fossils of most of the modern orders of mammals appear in a brief period during the early Eocene (55.5 - 33.7 million years ago). Both groups of modern hoofed animals, the Artiodactyla ("even-toed" taxa such as cows and pigs) and Perrisodactyla ("odd-toed" taxa, including the horses), became widespread throughout North America and Europe. The evolutionary history of the horses is particularly well understood: Stephen Jay Gould (1983) provides an excellent discussion of it in his book "Hens' teeth and horses' toes".

At the same time as the mammals were diversifying on land, they were also returning to the sea. The evolutionary transitions that led to the whales have been closely studied in recent years, with extensive fossil finds from India, Pakistan, and the Middle East. These fossils chronicle the change from the land-dwelling mesonychids, which are the likely ancestors of whales, through animals such as *Ambulocetus*, which was still a tetrapod but which also has such whale-like features as an ear capsule isolated from the rest of its skull, to the primitive whales called the Archaeocetes.

The trend towards a cooler global climate that occurred during the Oligocene epoch (33.7 - 23.8 million years ago) saw the appearance of the grasses, which were to extend into vast grasslands during the subsequent Miocene (23.8 - 5.3 million years ago). This change in vegetation drove the evolution of browsing animals, such as more modern horses, with teeth that could deal with the high silica content of the grasses. The cooling climate trend also affected the oceans, with a decline in the number of marine plankton and invertebrates. While DNA evidence suggests that the great apes evolved during the Oligocene, abundant fossils do not appear until the Miocene. Hominids, on the evolutionary line leading to humans, first appear in the fossil record in the Pliocene (5.3 - 1.8 million years ago). The story of human evolution is covered here - Human Evolution material.

New Zealand, by virtue of its isolation and its relatively recent geological development, was not the centre of any novel evolutionary development. However, many of the species that date back to Gondwanaland, or that arrived more recently as migrants, have undergone significant adaptive radiation in their new homeland. Some of the best examples of this can be related to the major ecological changes that accompanied the Pleistocene Ice Ages.

Throughout the Pleistocene there were about twenty cycles of cold glacial ("Ice Age") and warm interglacial periods at intervals of about 100,000 years. During the Ice Ages glaciers dominated the landscape, snow and ice extended into the lowlands, transporting huge quantities of rock with them. During these periods the South Island was extensively glaciated, and there were small glaciers on the Tararua Ranges and Central Plateau. Because a lot of water was locked up in ice, the sea levels dropped during the glacials (up to 135m lower than at present). Extensive land bridges joined the main and many offshore islands, allowing the migration of plants and animals. During the warmer periods large areas became submerged again under water. These repeated episodes of environmental fragmentation drove rapid adaptive radiation in many NZ species, especially (but not exclusively) the alpine plants.

For example, speciation patterns in the native *Placostylus* flax snails of Northland can be related to changes in sea level. Originally 2-3 species were widespread at a time of low sea levels. Rising seas at the end of the glacial period isolated these as populations on offshore islands, where differential natural selection pressures led to the evolution of a greater number of separate species.

The distribution of land snails such as *Powelliphanta* in Marlborough and the southern North Island also offers evidence for the presence of land bridges and the possibility of future speciation. The same varieties are found both north and south of Cook Strait, implying a continuous land bridge in the past as the animals die in salt water. The fact that no further speciation has occurred in this case suggests that the land bridge was recently submerged by rising seas, perhaps only 10,000 years ago.

The Correlation Between Bird Song and Human Speech

Bird Song and Human Speech

To learn their language, humans and white crowned sparrows follow similar steps, recognition, practice, and clarity. The best time for a human to learn their language is from toddler to twelve years old (Macdonald) and the best time for a white crowned sparrow to learn its song is between 10-50 days from when they were born. Both genes and environment play roles in determining the way each communicate (Alcock 25).

The first step for humans and white crowned species in learning their language can be considered recognition. White crowned sparrows and humans listen to a tutor before they begin to communicate their language. The tutor teaches the language and the particular dialect, according to the area. Just like humans, the sparrows also have a dialect depending on where they live. A white crowned sparrow has certain genes that only allow them to learn its own species' songs. Experiments have been performed in which the sparrow was kept in isolation and played tapes of other song sparrows, the white crowned sparrow will not imitate the other species' song, but will sing an odd song unlike either species' song (Alcock 24-25). Humans also cannot communicate with each other using a different species' language. A study was done in Avignon, France that observed children who were brought up by wolves. It was observed that thc children spoke no language at all. Before studies of this kind were done it was thought that the children might learn to communicate with the wolves, just like the Tarzan story. This is not true, the children could not speak the human language nor the language of the wolves (Macdonald). It is a combination of environment and genes that tell white crowned sparrows and humans to learn their proper language. Genes tell the species to only learn the language of their own species and environment plays a factor in determining which dialect each will use.

Once the white crowned sparrow or infant has recognised its species, song, they can begin to practice the language. Infants first communicate by making sounds. The sounds can range from grunts to sounds that mimic their surrounding environments. For example, before babies can say words, they might say moo- moo when looking at a cow. After sounds come words. By the time the infant reaches one year, they should be making sentences out of words (Macdonald). The white crowned sparrow also does not immediately master its song. The sparrow will first sing a short subsong derived from the tutor's full song. The sparrows keep practicing their subsong just as infants practice their words. After the sparrow has mastered its subsong it can start to form a full song. In both species it is necessary for them to hear themselves in order to vocalize their language correctly (Alcock 26).

In conclusion, both white crowned sparrows first listen and recognise their particular species' language. A white crowned sparrow has a tutor to teach the song. Infants usually have parents that teach them their verbal language. Then they must practice the language, starting off slow with sounds and building up. Both species will continue to improve and clarify their language throughout their lives.

Army Ants - Social Hierarchy, Life Cycles, and Migration

Army Ants (Eciton Burichelli)

Army ant colony castes consist of queen ants, soldier ants, and multi-purpose worker ants. In the army ant colony, the queen ant lays eggs, the soldiers focus their attention to the defense of the colony, and the worker ants are split between being foragers or tending to the queens brood. The Army ant is abundantly found in humid lowland forests or heavily forested areas of Central/South America. The changes in temperature and precipitation have no affect on the life cycles or migration of these ants.

The social structure and life cycles of army ants have two distinct phases. Army ant colonies go through a stationary phase and a migratory (nomadic) phase. The phases are based on the ability of a queen to produce many eggs in a very short period of time (from 100 000 to 300 000 eggs in five to ten days); thus, the queen overpopulates a given area, which then triggers migration of the colony.

In the stationary phase, the army ant queen's abdomen swells to hold 55 000 to 66 000 eggs. As the eggs develop into mature ants, tens of thousands of adult workers appear, triggering the nomadic phase: colony activity increases exponentially, and "swarm raids" increase in size and intensity. With a colony of this size, the army ant colony must emigrate daily. The migration continues until "larval pupation" begins. Once this occurs, the colony re-enters the stationary phase, and the cycle repeats. Due to their large colony sizes, army ants migrate in order to find food. Most ants are "column feeders." Army ants are what are called "swarm feeders." The foraging workers form a fan-shaped swarm with a broad front. Along with feeding, army ants also have a unique way of forming nests. The army ant nest is made up of army ants themselves. The ants form walls and fasten onto each other by using their mandibles. The ant colony forms these walls almost anywhere in order to enclose the queen and her brood. The nest is very structured, as it provides corridors within itself for transporting food and eggs throughout the many areas of the nest.

Army ant colonies march and nest in different phases as well. In the nomadic phase, army ants march at night and stop to camp during the day. The colony begins a nomadic phase when available food has decreased. During this phase, the colony makes temporary nests that are changed everyday. Each of these nomadic rampages or marches lasts for approximately 17 days. Army ants are notorious for eating anything that gets in their path. They can kill and eat up to 100 000 animals in a day! They normally consume lizards, snakes, chickens, pigs, goats, scorpions, tarantulas, beetles and other ants. Army ants can also climb trees, and feed on animals within its canopy.

Army ants communication is completely dependent on chemical messaging and trail pheromones. These methods of communication act as a stimuli for changing behaviour patterns. Unlike other ants, army ants do not have compound eyes, but instead have single eyes. Army ants are also blind and have to use their antennae to sense smell and touch. The army ants use these senses to communicate in nesting and raiding.

Mimicry and Camouflage

Mimicry

A major concern of animals and other critters is to protect themselves from predators in order to survive and reproduce and pass their genes off to a new generation. Many animals have evolved adaptations known as antipredator devices such as camouflage and chemical toxins. Animals use camouflage to blend in with their environments in an attempt to be unrecognisable by predators. Other organisms such as the monarch butterfly contain chemical toxins that are secreted into the predator's mouth when it attempts to eat the butterfly. The monarch butterfly also has warning colouration that gives a warning sign to predators to remind them that the butterfly is toxic and should not be eaten.

These antipredator devices are so successful that other organisms have been known to mimic them. The organism that is mimicked is known as the model and the third party that is deceived by the model and its mimic is known as the receiver. The mimics have learned to take advantage of the colour patterns and markings that predators have learned by experience to avoid. The model is usually a species that has an abundant population and has successfully warded off predators with an antipredator device.

Organisms have learned to mimic their surroundings or environment in an attempt to "hide" from predators. For example, lizards have learned to mimic tree trunk colour which proves to be very successful as predators will simply move past them as they believe that they are simply looking at a tree. Another example of this type of mimicry can be seen with the Katydid who will mimic a leaf in both colour and shape in an attempt to be hidden.

Some prey animals have evolved certain patterns on their bodies that mimic other animals in an attempt to startle their predators. The most common example of this type of mimicry can be found in some moths and butterflies who flash eye spots on their wings to predators. These eye spots startle the predator who believes that the eyes belong to a much larger animal that may be a threat to them.

In one form of mimicry known as aggressive mimicry, an organism will mimic a signal that is either deceptive or attractive to its prey. One example of this involves the praying mantis who will mimic flowers to attract insects that they can then capture and eat. Organisms can also imitate the behaviours of other organisms. Moth caterpillars, for example, will imitate the motion and body movements of a snake in order to scare off predators that are usually a prey item for snakes.

One of the most popular types of mimicry involves the warning colouration found on inedible or toxic organisms such as the monarch butterfly. Once these toxic organisms have adapted this warning colouration which warns predators to stay away, other organisms may start to mimic this warning colouration in an attempt to stay alive. Batesian mimics are those mimics that imitate unpalatable species even though they are palatable. Therefore, one species is harmful while the other is harmless. The wasp is a great example of Batesian mimicry. The wasp is the model species in this example as it possesses a sting which enables it to escape from predators. The bright warning colouration of the wasp has been mimicked by many other insects. Even though the mimics are harmless, the predator will avoid them due to bad experiences with wasps with the same colouration. With Müllerian mimicry, many unpalatable species share a similar colour pattern. Müllerian mimicry proves to be successful as the predator only has to be exposed to one of the species in order to learn to stay away from all the other species with the same warning colour patterns. The black and yellow striped bodies of social wasps, solitary digger wasps, and caterpillars of the cinnabar moths warn predators that the organism is inedible. This is a great example of Müllerian mimicry as all of these unpalatable, unrelated species have a shared colour pattern that keeps predators away.

Mimicry is a very successful antipredator device that species have evolved over many generations. As one can see, organisms have come to mimic many different characteristics such as colour patterns and behaviours. However, selection only favors the mimics when they are less common than the model. Therefore, the fitness of mimics is "negatively frequency-dependent."

Habitat

Space Utilisation

Individuals must find their place within the three dimensional expanse in which all material objects are located. Individual species ought to strive for finding the most favorable combination of spatial distribution and the use of resources offered there. Distribution and behaviour of individuals or species contingent on spatially explicit references.Habitat refers to any part of the biosphere where a species can successfully live and reproduce; Niche, in contrast, describes a species' ecological role within a community.

Individuals

- Ideal Free Distribution The relative number of foragers in each patch should match the relative, available amount of resources obtained there (except with incomplete information or inter-forager interference). Individuals should distribute themselves among habitats so that every individual maximises its net rate of return. Costs and benefits may depend on how many other users are already using the resource. applies to food, mates, etc.
- Ideal Depotic Distribution Individuals chose the most suitable area, within the constraints that other residents provide
- Allee-type Distribution Individuals chose the most suitable area, within the constraints that other residents provide

Fundamental vs. realised niche: A variety of factors may determine the presence or absence of a species at a particular site. Individuals pf a species may not have reached the habitat - dispersal. Alternatively, a variety of determinants may play a role, including abiotic (e.g., weather conditions, soil, climate) or biotic (interactions with other organisms, predation, competition, disease, social factors).

Strategies, Decisions and Preferences in behavioural Ecology: confer an adaptive advantage with success measured as <Fitness> Reproductive Success, the probability that an animal of a particular genotype and phenotype will manage to foster reproductively successful offspring.

Habitat Preferences

Natural selection will favourindividuals which utilise those habitats in which the greatest number of successful offspring can be raised. Spatial constraints may involve a balance of considerations: e.g., Herring gulls; Variables to consider: resource limitations; spatial and temporal distribution of the resource; variation in resource quality; number of competitors that attempt to control a resource; Predation pressure, etc.

- Habitat preference as a closed developmental system: many invertebrates, crayfish. Host plant preference in bladder grasshoppers. Poor response to rapid environmental change and human impact, e.g. Karner blue butterfly (Lycaeides melissa samuelis) depends on lupine (Lupinus perennis) as its sole larval food source. Species with highly fixed, genetically programmed habitat selection may require considerable time to evolve mechanisms for the selection of a new habitat
- Habitat preference as an open developmental system:
  - habitat imprinting: many migratory vertebrates, such as salmon tradition: mountain sheep, greylag goose migration
- Dispersal Ecological process affecting distribution, and genetic process affecting geographic differentiation and variation; leaving an area of birth or activity; largely a short range effort; Colonisation - long range dispersal. Stay in the safety of home: little danger but competition may reduce reproduction or Disperse and take chances: high mortality but high payoffs. Natural dispersal: e.g. Cattle egrets or Intended and unintended aid in dispersal by humans: e.g. European starling, house sparrow, Africanized honeybees
- Descriptives for Spatial Distributions: Are distributions random, clumped, dispersed? Describe distribution with: minimum polygon, bivariate normal, spatial autocorrelation, contiguous clustering, or non-parametric utilisation distribution (UD).

Chapter 3

Hormones in Animal Behaviour

Hormones

The Glandular Systems

The body has two types of glandular systems, the endocrine, which generally secrete hormones through the bloodstream, and the exocrene which secrete fluids to the outer surfaces of the body, such as sweating.

Endocrine Glands

The Endocrine System combines neural and glandular mechanisms which control physiological functions/ behaviour via the secretion of hormones. Hormones are chemical signaling molecules which play an integral role during development (organisational effects) and day-to-day functioning (activational effects) of target tissues at critical times. Secretory cells of a particular type are often clumped together into a well defined gland (e.g. pituitary, thyroid, adrenal, testes, ovaries). Secreted at that site they distribute throughout the body via the blood stream, and cause physiological changes at any other sites.

- Steroids: derived from cholesterol
- Amines: derived from amino acids
- Peptides: short chains of amino acids via protein synthesis.

Sex hormones, largely steroids, are secreted from gonads and adrenal cortex. Androgens (e.g., testosterone) are usually higher in male mammals while levels of estrogens (e.g., estradiol) in female mammals exceed those in males. Circulating levels of sex hormones then provide the basic organisation for gender phenotypes.

Neurohormone refers to a compound that is released into the bloodstream at specialised neurohemal release sites. It binds to receptors anywhere in the body and thereby coordinates disparate biochemical responses.They are released from glands, transported via the circulatory system and influence the activity of target organs.

Functionally hormones are categorised as Effector hormones (e.g. Vasopressin, Oxytocin) or Tropic hormones, releasing factors (e.g. Gonadotropin Releasing Hormone - GnRH, Growth Hormone Releasing Hormone - GHRH). Target Organs receive hormones via blood stream, respond directly or release their own hormones in response (steroid hormones), and these hormones circulate back to turn off hormonal secretion: endocrine feed back loops. Compensatory hypertrophy results from feedback loops control levels of activity where systems are upregulated until they achieve sufficient functional effects (e.g. thyroxine and goiter). Activational Effects and Organisational Effects

Sex Determination

Hormonal influences during critical periods produce fairly permanent changes in nervous and endocrine systems. Sex-determination may be chromosomal or environmental:

- Mammals: hormonal cascade of events takes an individual down one of two paths (default is female). testes determining factor (tdf) in mammals or Sexchromosomal Abnormalities: Turner syndrome (XO monosomy), Klinefelter syndrome (XXY trisomy), XYY syndrome ("supermale"), Multi-X syndrome ("superfemale"), XX Male syndrome (SRY gene transference) XY Female syndrome (SRY gene missing)
- Birds: Males are homogametic (ZZ), females are heterogametic (ZW)
- Turtles: temperature
- Fish: social stimuli.

Internal sex organs: Precursors for both internal sex organs are present in the embryo: the Mullerian system for the female sex; the Wolffian system for males. Controlled by the levels of circulating hormones, only one set of organs develops while the other shrinks. Secreted by the testes, Müllerian inhibiting hormone (i.e., a peptide hormone) has defeminizing effects while androgens (i.e., steroid hormones) exert a masculinizing effect. Gonads: testes or ovaries, first to develop. External genitalia: Androgens are essential for the development of primary sexual characteristics.

Androgen Insensitivity Syndrome: A genetic male lacking functional androgen receptors develops testes, secretes Müllerian inhibiting substance and androgens. Internal sex organs progress onto the intended male path, but female external genitalia result from the inability to respond to androgens correctly. Turner's syndrome, with genotype XO (i.e., a single X chromosome and no Y), lacks testes and ovaries.

Maternal effects explain influenced by effects of the mother that are not due to direct genetic inheritance. Positional effects in the mammalian uterus

Homeostasis

Hormonal secretion or its inhibition may bring about some rapid physiological changes in a variety of target tissues.

Behaviour

Hormones influence sensory perception (human visual, rat odor, preference in castrated vs. intact). Hypothalamus neurons monitor the internal state (thirst, hunger) and send neurosecretory cell axons to the posterior pituitary. In females, Oxytocin, released from the posterior pituitary, triggers milk let-down in mammary glands or contraction in the uterus during child-birth. It is also associated with powerful emotional effects (affective states) in parental behaviour and maternal competence.

Moreover, the hypothalamus secretes releasing factors to stimulate or inhibit manufacture and secretion of hormones in anterior pituitary portal system: Prolactin, Follicle stimulating hormone (FSH), Luteinizing hormone (LH); receives, stores and releases neurohormones from hypothalamus. Sex behaviours: gonadotropin releasing hormone (GnRH) -> gonadotropin -> gonad maturation and gonadal steroid production -> feed back to brain. Sex steroids initiate the basic female and male anatomies (i.e., organising effects in sexual differentiation cascade) but also allow animals to later respond to activational effects. Expression of sex differences in behaviour also requires that steroids activate many aspects of the phenotype during maturation.

Reproductive Neuroendocrinology of Ring Doves (*Streptopelia risoria*): monogamous, sexually monomorphic, parental care from both male and female (incubate eggs, crop "milk"); Lordosis in female hamsters: female soliciting behaviour. Lee-Boot effect - estrous cycles slow and stop in female mice housed together; Whitten effect - individuals start cycling again in synchrony if exposed to male odor; Vandenbergh effect - acceleration of onset of puberty in female rat when exposed to odor of male; Bruce effect - failure of recent pregnancy of female rat when exposed to male who is not the father.

Arthropod Molting

Hormonal Control of Arthropod Molting: Brain: releases ecdysiotropin which stimulates release of molting hormone (MH) from Prothoracic Gland; Corpora allata: Juvenile Hormone (JH) suppresses metamorphosis; Prothoracic Gland; Corpora allata: Juvenile Hormone (JH) suppresses metamorphosis

Honeybee Foraging Behavioural Analysis

The common honeybee (*Apis mellifera*) forages around the hive in search of resources. Usually this resource constitutes food in the form of pollinating flowers. Several landmark studies were performed in the middle of the 20th century to analyse how these social organisms communicated exact distances and direction to one another in order to effectively locate these resources. Karl von Frisch determined that honeybees perform two distinct dance routines that coincide with two different distance approximations made by the foraging bee. These two dances, the Round dance and the Waggle dance, communicate to the other the approximate distance from the hive to the new resource. Only the Waggle dance communicates direction (von Frisch 1967).

The Round dance is performed by the returning bee usually in complete darkness, vertically on a honeycomb. The circuitous motion attracts other foragers, which then learn that the resource is within approximately 50 meters of the hive. No direction is given by this routine (von Frisch). As a result, the newest foragers leave to search in all directions surrounding the hive. behaviourally, this dance is energetically favorable due to the short distances travelled. In contrast, the Waggle dance is energetically unfavorable to the individual, but beneficial to the hive.

The Waggle dance is performed primarily when the resource is further than 50 meters. The returning forager either performs the dance on a vertical surface or a horizontal one. To determine distance and direction, the bee orientates itself relative to the sun. Any deviation from this point gives the angle the new foragers should pursue. If vertical, the bee orientates itself to gravity. Perpendicular to the ground becomes the reference point (i.e. the sun). Deviations from such relay direction accordingly. Distance is communicated by the length of the abdomen shake that forms the middle of a figure eight dance (von Frisch). Kirchner et al. determined that the Round dance does not convey direction. By tipping several hives onto their horizontal axis and placed a resource 10 meters away, the reference heading was experimentally altered. Regardless of the hive orientation, the number of successful returning foragers was constant (1998). Once the distance crossed 50 meters, the success rate decreased for hives orientated horizontally, indicating direction is dependent on hive orientation for the waggle dance to be effective.

Honeybees foraging behaviour can be interpreted from a historical or adaptationist view. Lindauer performed historical analysis that reveals that distantly related bees might have evolved through 3 stages of development: First stage analysis reveals that the genus *Trigona* conveyed direction by buzzing to gain their hivemates attention. The odors trapped on the bee trigger others to forage in search of the odor. The secondary stage involves marking the path from the resource to the hive with mandibular pheromones. She then buzzes, and the bees follow the scent markers. The third stage involves an in-flight "waggle dance" in the direction of the resource (Lindauer 1961).

An adaptationist theory on bee dance evolution. The foragers seek new resources to provide increased hive fitness. By dancing, the cost/ benfit ratio is reduced to the individual level more so than a group of foragers expending energy searching in all directions all the time (Seely et al. 1998). Their study demonstrates that dancing has become an evolutionarily desirable trait because less time foraging allows for more time collecting the resource. Honeybee dancing provides an evolutionarily advantageous behaviour which optimises the hive's fitness.

Scientific Method

The Scientific method refers to a body of techniques for investigating phenomena, acquiring new knowledge, or correcting and integrating previous knowledge. To be termed scientific, a method of inquiry must be based on gathering observable, empirical and measurable evidence subject to specific principles of reasoning. A scientific method consists of the collection of data through observation and experimentation, and the formulation and testing of hypotheses

History

Socrates developed logic as a way of thinking and speaking which would let you prove that a certain statement was or was not true. Plato (Socrates' student) continued this idea, and Plato's student Aristotle began to apply logic to the natural world in which arguments were used instead of experiments. Aristotle, dissatisfied with the knowledge that was imparted when the researcher moves from a set of specific facts to a general conclusion, rather relied on deductive reasoning. For centuries the library of Alexandria served as a center for scientific exploration. Following Roman conversion to Christianity its influence in the Western world declined around 600AD but its contents was preserved with the rise of islam where explorations of the natural world prospered. The experimental scientific method was developed by Alhazen, who in 1021AD used experimentation and mathematics to obtain the results in his Book of Optics. In the medieval Western world (600-1000) faith and the supernatural reigned supreme over the value of human reason and science. Giving rise to the Renaissance, the grip of church doctrine weakened after the 12th century with a variety of influences, including the black death, the influx of scholarly literature from the islamic library of Cordoba, and the printing press. Galileo Galilei (1564-1642) starts the Scientific Revolution and claims that church doctrine ought not to make claims about the natural world which can easily be shown to be false - and he pays the price. Renee Descartes (1596-1650) urges us to never accept anything as true that is not known clearly to be such. William Harvey (1578-1657) phrases the Scientific Method.

A Method for Knowing

The scientific method is the process by which scientists construct an accurate (i.e., reliable, consistent and non-arbitrary) representation of the world. Scientists attempt to minimise the influence of their biases on the outcome of an experiment. The most fundamental error is to mistake the hypothesis for an explanation of a phenomenon, without performing experimental tests. Sometimes "common sense" and "logic" tempt us into believing that no test is needed. There are numerous examples of this, dating from the Greek philosophers to the present day.The prime directive is to never accept anything as true that is not known clearly to be such. Towards this goal scientists progress through a series of structured steps.

- Initial Observation tickles your interest in a particular topic
- Phrase a Question that can be addressed with an experiment
- Phrase Null Hypothesis as an explicit claim about the value of a specific parameter. What would you predict if the decision to turn left/right is random? In our example, the probability to turn right P(R) must equal the probability of turning left P(L). Because subjects have two mutually exclusive choices at the decision point we know that the probability of turning either left or right P(R or L) = P(R) + P(L) = 1, or P(R) = P(L) = 0.5. Thus, we can make an explicit claim about the probability that a subject turns right or left at the decision point given the choice is random. Our null hypothesis is that P(R) = 0.5 (or, equivalently, that P(L) = 0.5).

  - Directional vs. non-directional hypotheses
  - Rejection regions and one-tailed vs. two-tailed tests
- Phrase Alternate Hypothesis which simply states that our null hypothesis is not true, or P(R) 0.5 (or, equivalently, P(L) 0.5).
- Consider issues relating to the Treatment Effect
  - Decide what size of a treatment effect would be considered "significant"
  - Decide on sample size needed to detect a treatment effect of this size (i.e., power)
- Decide on framework (experiment) for distinguishing between the two hypotheses. This allows us to attach a sense of likelihood to the outcomes associated with a particular outcome under the null hypothesis. Scientists must minimise their own biases with respect to a specific outcome and their preferences may not bias the results or their interpretation.
- Consider the type of data that the experiment will produce and choose a statistical test that allows you to assess to what degree the experiment's outcome may be the result of chance alone. This step involves the computation of a specific test statistic. You are bound to retain the null hypothesis unless you have good evidence that it is not compatible with the outcome of the experiment. A comparison of the test statistic to a critical value allows us to either retain our null hypothesis or reject it in favourof the alternative hypothesis.
- Choose acceptable level for falsely rejecting the null hypothesis (i.e., p-value)
- Determine critical values for the sample statistic. We accept the null hypothesis if our test statistic does not exceed the critical value and we reject the null hypothesis in favourof the alternative if the test statistic exceeds the critical value.
- Conduct Experiment in randomised, double-blind fashion with the inclusion of appropriate controls
- Un-blind samples. Match each particular outcome with their respective treatments
- During Data Analysis all data must be handled in the same way. There is always the temptation to more closely inspect the validity of data points which do not support the scientist's expectations, while data which do agree with those expectations may not be checked as carefully.

- Draw Conclusions
- Communicate Results.

Terminology

The general public often critically misinterprets terms used by scientists such as hypothesis, theory and scientific law. A hypothesis is any educated hunch phrased to explain an observed phenomenon so it becomes testable by an experiment. Even the most hair-brained ideas can be phrased into a hypothesis. Model or principle is reserved for situations that have at least limited validity. For example, Bohr's model of the atom is formed as an analogy to the solar system, where electrons move in circular orbits around the nucleus.

This is not an accurate depiction of what an atom "looks like," but the model is to some degree able to represent the energies of electrons in a hydrogen atom. The term Scientific Law is reserved for a concise explanation of a simple set of actions that is accepted to be true and universal. Scientific laws are similar to mathematical postulates. The latter really no longer need complex external proofs or experimental verification based upon the fact that they have always been observed to be true.

Examples include the law of gravity, or the laws of thermodynamics. Both, laws and theories can be used to make predictions about the outcome of future events. Like a law, a theory is not any hair-brained idea that popped into somebody's head - it is rather an understanding of something that has been well documented, supported by overwhelming scientific evidence, has stood up to exhaustive rigorous testing, and is generally accepted as being true beyond reasonable doubt. Whereas laws govern a single action, a series of related phenomena can only be represented by a theory that describes the fundamental properties and relationships in a complex system. As with an automobile used as a transportationd device, improvements are sometimes made to one or more component parts (e.g., a new set of spark plugs), but the function of the automobile as a whole remains unchanged. Similarly, individual components of a theory can be changed or improved upon, without changing the overall truth of the theory as a whole. Examples of scientific theories include the theory of evolution, the theory of relativity, and quantum theory. Although scientists continue to attempt to make its components more elegant, concise, or all-encompassing, they are seldom, if ever, entirely replaced.

Ockham's Razor: "Pluralitas non est ponenda sine neccesitate, *which translates as "entities should not be multiplied unnecessarily —*

William of Ockham, 14th century. It reminds us to keep things simple. In cases where we have competing theories describing the same phenomenon we are bound to go with the most parsimoneous explanation first. It does not mean that the simplest theory will be correct, it just focuses on priorities.

Tautological Arguments are a needless repetition of an idea in different words or a logical statement that is necessarily true because it includes all possibilities. It is logically true irrespective of whether the underlying statement is factually true or false because no alternate hypothesis is possible.

- Competitive Exclusion Principle: When two ecologically identical species compete, one will exclude the other or they will coexist
- Natural Selection: If fitness is defined as survival. Which one survives? - "The fittest". Who are they? - Those that survive.

Circular Arguments assume the very thing it aims to prove. Statements may be logical consistent, but they do nothing to convince one of the truthfulness of the speaker. We take it for granted that proposition a implies proposition b. If we suppose that Proposotion a is correct, then proposition b has to be correct, right?

- "The Bible is the inspired Word of God because the Bible says so"
- "My arguments are correct because I am always right"

Biological Rhythms

Our surrounding undergoes predictable, cyclical fluctations as a result of changes in seasons, or the time of day. When events and conditions repeat in a rhythmic pattern, the ability to predict, anticipate, and prepare for them is a highly beneficial trait. Not surprisingly, most animals are able to adjust their physiology, behaviour, and life cycles to the upcoming conditions. Owls begin to stir in the late afternoon and evening, awaiting the emergence of small mammals at dusk. Ground squirrels gather rations and pack on fat reserves in the fall in preparation for cold winters spent underground. Moose reproductive cycles match the birth of fawns in the spring to the rich emergence of forage at that time. Human core body temperature cycles with a low during the middle of their sleep cycle and highs around lunch time and early evening. Bioluminescent fireworm of Bermuda time their romantic gettogethers to diurnal as well as lunar cycles. Adult amergence of 17-year cicadas, or bamboo flowering every few decades, is closely timed across entire populations. Chronobiology, the science that describes timing in biological clocks and their associated rhythms, makes extensive use of terms and concepts derived from engineering disciplines. Rhythmic events are described as to their period (i.e., the amount of time it takes to go from peak to peak), frequency (i.e., the number of cycles completed within a specific unit of time), amplitude (i.e., the distance between peaks and average), and phase (i.e., the timing of the rhythm relative to some objective, external point in time).

The timing of a cycle's events can operate at many different time frames. Some cycles are link to obvious external patterns including:

- sun: circadian refers to 24h period length, ultradian to cycles exceeding 24h
- moon: lunar cycles, lasting 24.8h, are controlled by the phase of the moon, while tidal rhythms of 12.4h are determined by the timing of its gravitation pull
- tilt of earth's spin axis: a circannual rhythm spans events that repeat on a 12 month basis corresponding to the Earth's seasons.

Biological Clocks

Many observed rhythms in physiology and behaviour often crucially depend on the presence of endogenous cycles and their production through biological clocks. Periodic rhythms, which are not simply responses to external periodic cues, have been documented for most living beings, including bacteria, fungi, plants, and animals. Biological clocks are self sustaining oscillators which will continue a period of free-running cycling even in the absence of external cues. However, clocks are usually linked to and can be reset by the environment via cues (i.e., Zeitgeber). Such entrainment keeps an organisms clock synched to its surrounding conditions.

Circadian clocks must contain a minimum of three basic elements: (1) input pathway(s) relay environmental information to a circadian pacemaker; (2) the endogenous pacemaker (oscillator) generates temporal patterns; and (3) output pathway(s) for the pacemaker to regulate output rhythms. A search for biological clocks commonly attempts to isolate an organism from its external cues and to search for the continuation of rhythmic patterns. Isolation from all possible (e.g., geophysical, magnetic, or radiation) cues remains a difficult task, however, animals will display cyclical activity patterns even when maintained in constant conditions aboard a spacecraft orbiting far above the earth.

Genetic components of the biological clock have emerged from research on the fruit fly where mutant lines of flies displayed abnormal cycling, including a shorter period, a longer one and its complete absence in yet another. All three mutations were mapped to the same gene - termed period or *per*. Disruption of the same gene in human sleep disorder underscored the conserved nature of the molecular circadian clock throughout evolution.

Biological Clocks exhibit a high degree of inheritance, independence of temperature and social conditions, strong resistance to pharmacologcial and chemical disruption, and may even be expressed at the level of single cells. Entrainment is often limited to a narrow range of possible outcomes, may utilise a small set of cues, and its timing is critical. Human circadian rhythms free-run at slightly longer than 24 hours and are reset to the normal 24h day by light cues. Humans are able to readily adjust to a 23-25h day but not to a 22 or 28h day. Biological rhythms may be expressed over many different time frames, and entrainment of one clock may disrupt multiple rhythmic changes suggesting a basic linkage of patterns (e.g., circadian activity and estrous cycles in mammals).

Karl von Frisch's search for biological clocks characterised a periodic, night-time lightening of skin chrmatophores in minnows. These cyclical changes in skin pigments persisted even in blinded individuals. With damage to the pineal gland, however, fish were no longer able to change skin colour in rhythmic fashion. The existence of extra ocular light perception in the brain's pineal gland allowed the fish to change colour even when ocular light perception was disrupted.

Molecular Basis of Cellular Clocks

The presence of rhythmic patterns even at the level of single cells indicates the presence of cell-autonomous circadian mechanisms. In the fruit fly (*Drosophila*) two proteins (PER and TIM) play a role in rhythmic patterns at the cellular level. The two proteins form a dimer complex in the nucleu, binds to the gene's promoter, and inhibits production of further *per* and *tim* RNA. With gene transcription of the component proteins turned way down, few new PER and TIM proteins are made. Existng PER-TIM complexes are gradually degraded, and decrease in numbers as they are not replaced. The protein-induced inhibition of *per* and *tim* transcription weakens, mRNAs for them increase, leading to enhanced PER and TIM protein, complex formation between them, and renewed inhibition of *per* and *tim* gene transcription. In the mammalian suprachiasmatic nucleus entrainment derives from light-induced expression of transcription factors (i.e., immediate early genes, IEG), which control the expression of timing genes down-stream. At the output level, the circadian clock controls the transcription of a number of genes such as CREM (i.e., CRE modulator).

Suprachiasmatic Nucleus

The mammalian circadian system relies on oscillating neurons grouped within the suprachiasmatic nucleus (SCN), a distinct group of cells located in the hypothalamus. Its destruction results in the complete obliteration of regular sleep/wake rhythms. Cultured SCN cells maintain their own rhythm in the absence of external cues. The SCN receives information on day length from the retina, interprets it, and during the dark phase enhances secretion of the hormone melatonin from the pineal gland.

Sleep

Sleep, an essential state of natural rest, is observed in most animals. It is characterised by reduced voluntary body movement, decreased reaction to external stimuli, and a loss of consciousness. Humans sleep an average of 7.5 hours, during which they proceed in around 100 minute cycles of two broad types. Rapid eye movement (REM) occupies roughly 25% of sleep time. During this stage the body shows a loss of skeletal muscle tone while the brain is quite active. Electroencephalography (EEG) shows a mixture of frequencies that is similar in appearance to the wakeful EEG and coincides with periods of intense dreaming. Non-rapid eye movement (NREM) sleep occupies most of the sleep time and is characterised by frequent limb movements and sleep walking while the brain exhibits reduced activity and little dreaming.

The inherent function of sleep is not yet completely understood. Sleep, which affects the immune system and metabolism, supports a restorative function with increases in wound healing and the production of anabolic growth hormones. Sleeps is marked by a variety of physiological processes of growth and rejuvenation of the organism's immune, nervous, muscular, and skeletal systems. In newborns sleep appears to play a critical role in establishing proper functional connectivity in the brain. A number of studies suggest the existence of a correlation between sleep and the many complex functions of memory. Memory and learning involve nerve cell dendrites sending information to the cell body to be organised into new neuronal connections. Sleep which reprocesses the days events in the absence of external information is processed by these dendrites as memories are solidified and knowledge is organised [Saper and Stickgold 2005].

Population Genetics

Population Genetics and Behaviour

Population genetics described the statistical distribution of genes in a particular breeding population, such as a breed of dog, and examines the role of various biological effects on gene distribution. While ordinary genetics predicts phenotypic makups of the next generation, population genetics concerns itself with genetics predictions for a breeding group several generations into the future.

Population

A population is the collection of organisms of a particular species within a given geographic area. When all individuals in a population are identical with regard to a particular phenotypic trait they are known as monomorphic. When the individuals show several variants of a particular trait they are polymorphic. The Gene pool for a species or population is the complete set of unique alleles present in it. While large gene pools indicate robust populations, extensive genetic diversity, and a good chance to survive bouts of intense selection, low genetic diversity is associated with reduced biological fitness and an increased chance of extinction.

Hardy-Weinberg Equilibrium

Hardy-Weinberg Equilibrium: Allele frequencies will remain constant across generations if the population is large and individuals are mating randomly. Moreover, we will be able to accurately predict genotype frequencies from gene frequencies (e.g., in a two-allele system - a2+2aA+A2). HW produces the null hypothesis against which genotype frequencies are compared. HW equilibrium will occur after one generation provided that the same gene frequencies occur in both sexes. If disequilibrium occurs, equilibrium will be re-established after one generation of random mating. What are the assumptions? What will produce deviations from the HW equilibrium? What might produce an excess of heterozygotes (e.g. HLA-1, MHC)? What might produce an excess of homozygotes? Deviations from HW equilibrium mayt derive from a variety of sources.

Mutations

Mutations are alterations in genetic code ranging from changes to individual nucleotide bases along the DNA to largescale rearrangements of chromosomes. Mutations are rare and even at maximum rates allele frequencies will only change from p=1 to p=0.9993 per generation.

Genetic Drift

With small effective population size, random events can take on increasingly prominent roles. This focuses on changes in the genetics of the population as a result of chance alone.

- Bottlenecks
- Founder Effect

Examples include the relatively high rate of an autosomal recessive skeletal dysplasia (Ellis-van Creveld Syndrome) in the Amish population of Lancaster, Pa. Resulting in short-limbed disproportionate dwarfism, its incidence within the population is approximately 5 per 1000 live births and 2 per 1000 living persons. It can be traced to the initial immigrants Samuel King and his wife.

Selection

Selection: Strong selection for traits with high heritability can change proportions in excess of 10% per generation

Non-random Mating

Non-random Mating: Mate Choice with assortative mating (e.g., blue and white form of snow geese), Positive assortative mating: mating among individuals who share particular genes or phenotypes. Negative assortative mating: preference for mating among individuals who are unlike each other with respect to particular genes or phenotypes. Inbreeding and Outbreeding. Inbreeding: refers to (preferential) mating between biological relatives. As relatives (i.e., ancestors of the first individuals are shared with those of the second individual), they carry genes which are "identical by descent"; extreme inbreeding: mating between sibs, half-sibs, parent-offspring. Individual inbreeding coefficient (i.e., Pedigree inbreeding): F represents the probability that the offspring is homozygous due to identity by descent (ibd) at a randomly chosen autosomal locus, ranges in value from 0 (no locus ibd) to 1 (all loci ibd).

Significant factors for pedigfree inbreeding are Factors which determine the probability that various kinds of individuals will come into contact, e.g., population demography (sex ratio, birth or death rates), individual and population dispersal patterns, spatial distributions behavioural preferences for or against certain classes of individuals as potential mates. behavioural preferences influence the acceptance or rejection of different types of individuals as mating partners:

- Population inbreeding coefficient: f measured by deviations from Hardy-Weinberg Equilibrium (a2+2aA+A2), examine for an over-representation of homozygotes. f ranges from -1 (inbreeding avoidance) to 0 (random mating) to 1 (inbreeding)
- Outbreeding: (preferential) mating between non-relatives
- Inbreeding depression: overrepresentation of homozygotes; e.g. matings between cousin are rare (0.05% in U.S.) but account for 20% of albinos; deficiency in immune systems; recessive, deleterious alleles occur more likely in homozygous condition;

humans have on average 8 recessive, deleterious alleles. 1% increase in inbreeding produces 11% increase in the incidence of genetic diseases; Even small amounts of pedigree inbreeding in a random mating population greatly increase the incidence of some types of genetic disease. This is of particular concern in groups with small effective population sizes - Amish, Ashkenzi Jews

- Inbreeding advantages: functional association of gene complexes shaped by local adaptations
- Outbreeding depression: Extreme case, hybridisation across genetically differentiated populations
- Hybrid vigor.

Sociality

Cooperation, Altruism, and Sociality

Many animals and humans show a startling flair for exhibiting kindness, cooperation, and altruistic behaviour. Evolution, which is often (and falsely) viewed as largely driven by conflict between competing individuals, may not be an obvious candidates for explaining their existence. Evolution may promote traits underlying group living, altruism, and coorperative behaviours but only if genes underlying such characteristics are favored.

- Cooperation refers to the practice of individuals working together instead of working separately in competition. The success of participating individuals depends and is contingent upon the success of their partners in cooperation.
- Altruism consists of unselfish/selfless acts in which the altruist appears to put other individual's interests before their own. The general expectation is that the evolution of altruistic traits is not compatible with selfish gene theory. How could such a mechanism reinforce a behaviour jeopardizing one's own interests at the expense of another's? Examples include self-sacrificing behaviour: e.g. vampire bats, guard bees, Cooperative breeding, e.g. African Kingfishers, Acorn Woodpeckers, Florida Scrub Jay;
- Social behaviour relates to interactions with other members of the population. Group living comes with a diverse range of advantages and disadvantages. A Society describes a group of individuals of the same species, organised in a cooperative manner, and extending beyond sexual contexts and parental care.

- Eusociality is an extreme form of altruism where sterile workers labouron behalf of reproductive individuals, cooperate in the care of young. There must be an overlap of at least two generations of individuals which share in colony labor.

Units of Selection

Discussion about the units of selection centers on an understanding of what the life of each individual is ultimately devoted to. This is nowhere more pertinent than in evolutionary explanations for altruistic acts or for the emergence of highly social societies.

- Group Selection: Altruistic behaviour could be rooted through group selectionist thinking where altruistic traits favourreproductive success of a group at the expense of that of an individual (Wynne-Edwards). Groups containing altruists may thereby be able to outcompeting other groups, which lack such cooperation. Unfortunately, all members of a group displaying a high level of cooperation will derive the benefits, even those individuals who may be less cooperative than others. "For the good of the species..." explanations have not fared well in theoretical or empirical studies.
- Individual Selection: If individuals are simply looking out for their selfish interests, how could altruistic behaviour ever emerge as a superior strategy to it?
- Gene Selection: Is altruistic behaviour focused on the preservation of selfish genes

Simple Egoistic Solutions

In some instances, benefits may accrue to an individual committing an altruistic act. If these benefits exceed its cost, it is in the egoistic interest of the individual to perform such an altruistic act.

Synergistic Benefits

Underwater video (looping) of a school of Atlantic herring (*Clupea harengus*) on its migration to their spawning grounds in the Baltic Sea. Fish may derive many benefits from shoaling behaviour including defense against predators (by enhanced predator detection, distraction during prey strike, and diluting an individual's chance of capture), enhanced foraging success, higher success of finding a mate, and increased hydrodynamic efficiency.

Cooperation may be in each individual's egoistic interest if it increases its own individual fitness. Wolves, for instance, hunt in packs and then share their prey. Thereby each wolve benefits from pack hunting as the group can bring down prey that would be too large for a single individual. Although the behaviour may seem altruistic, individuals benefit directly in tasks that require many eyes. Fish tend to group in schools as it becomes significantly harder for a predator to pick out and hunt one specific individual from within their midst.

In 1971, Hamilton proposed the Selfish Herd model, which postulated a centripetal instinct that directs animals to place themselves towards the centre of a group. This behaviour is based on a presumed increase in predation towards the outer edges of the group.

Delayed Benefits

Young may remain with their parents and help them or group members raise more offspring, rather than breeding themselves (e.g., ostriches, some primates). Helpers may gain useful experience in raising their own offspring or have an increased chance to inherit a valuable breeding territory.

Reciprocity

Reciprocal altruism exists in situations where individuals exhibit altruistic acts towards each other. For each individual the benefits accrued must exceed the costs it incurs.

Complex Egoistic Solutions

For Individuals: The Prisoner's Dilemma: e.g. Pukekos (*Porphyrio porphyrio*).

True Altruism

An individual behaves in such a way as to enhance the reproduction of another individual, at a cost to its own fitness. (e.g., Sterile workers in social insects who give up all reproduction for the benefit of their mother queen. Mutual or delayed benefits can't account for this one: sterile workers never get to produce any daughters.)

Inclusive Fitness

An individual fitness solution to the paradox goes as follows: genes are favored that produce a disproportionate propagation of ones genes to subsequent generations even if it is at the expense of the individual. This may occur directly through either reproducing personally or indirectly by encouraging the reproduction of close relatives who share many of ones genes. Under those circumstances altruistic behaviours should primarily be directed towards close relatives who share many of the genes as compared to non-relatives who share fewer of them. Inclusive fitness refers to the sum of an individual's *Direct Fitness* (probability of reproductive success of one's own offspring) and its *Indirect Fitness* (probability of reproductive success of non-descendant relatives). Altruistic acts are frequently directed towards relatives. <Kin Selection>: selection for traits that lower an individual's personal fitness, but raise a relative's fitness. Recipient (related) kin share genes with the altruistic individual and are thus genetic extensions of them. Kin selection and inclusive fitness can explain sociality through increased survivorship of relatives, increased inclusive fitness, altruism, and delayed maturation or breeding.

Inclusive Fitness: Refers to the sum of an individual's <Direct Fitness> (probability of reproductive success of one's own offspring) and its <Indirect Fitness> (probability of reproductive success of non-descendant relatives). Altruistic acts are frequently directed towards relatives. <Kin Selection>: selection for traits that lower an individual's personal fitness, but raise a relative's fitness. Recipient (related) kin share genes with the altruistic individual and are thus genetic extensions of them. Kin selection and inclusive fitness can explain sociality through increased survivorship of relatives, increased inclusive fitness, altruism, and delayed maturation or breeding. Kin selection favors behaviour that sufficiently enhances the fitness of related individuals (i.e., carrying similar genes) leading to an increased representation of genes being passed on to the next generation.

Hamilton's Rule benefit to recipient / cost to altruist > 1 / coefficient of relatedness.

Consequences

The complex needs for group-living may have driven members towards a surprising degree of intelligence. Individuals of many species are able to keep track of each other's social status, correctly attribute each others mental states, and successfully engage in each other's deception.

Modeling

Modeling Adaptive Behaviour

It is a daunting (or impossible) task to study behaviour by simultaneously considering all significant factors and interactions that may impact it. An alternative method attempts to reduce the associated complexities by wrapping ones exptectations into a set of mathematical abstractions based on a reduced number of significant variables. Resulting quantitative predictions are necessarily incomplete but are often less ambiguous and easier to test. The goal is to model an animals rational decisions of what action to take, given some information about the world. The player then faces consequences for the decision as a function of the action, the actions of others (if applicable), and the state of the surrounding environment. The player is expected to be rational by maximising the expected utility.

Optimality Models

Optimality models quantitatively predict the consequences for a particular animal to behave in a certain way when pitted against the environment. It assumes that animals should behave so as to maximise their fitness. Optimality models consider separate, dependent, fitness benefits (B) and fitness costs (C) over a given range of values in a decision variable (i.e., the variable of interest). The solution to an optimality model attempts to find the point where the net benefits (i.e., B-C) are maximised. Individual solutions consider the consequences for a focal animal regardless of what strategies other individuals are relying on.

The main advantages of optimality models are in that they:

- help us clarify our assumptions
- force us to think about the information the animal has
- tie many different ideas together in one place
- generate testable predictions.

Consequences of behavioural decisions are described as equations where the success of decisions of one individual do not depend on what decisions other animals make.

- Decision variable is the variable that we aim to optimise (e.g., optimise speed to obtain the best mileage when driving from here to there)
- Currency refers to the criterion used to examine the outcome for different values of the decision variable (mileage)
- Constraints are factors that limit the relationship between decision variable and currency (e.g., speed limits).

Optimality models use calculus and logic to solve for a minimum or a maximum in a specific function f(x). This can be achieved by taking the derivative of the function df[x)/dt (i.e., to obtain the slope of the function) and setting it to 0. Such an approach can also be used to explore sensitivity to changes in values of the decision variable and to determine what types of information are relevant for the relationship. Precise predictions can be formalised for empirical testing

Ideal Free Distribution

An ideal free distribution describes the way in which animals distribute themselves among several patches of resources. The theory states that individual animals will aggregate in various patches proportionately to the amount of resources available in each. So for example, if patch A contains twice as much food as patch B, there will be twice as many individuals foraging in patch A as in patch B.

Predator - Prey Population Cycles

In order to study predation and population oscillations, Huffaker used mite species, one being the predator and the other being the prey. By modifying the spatial structure of the habitat, he could manipulate the population dynamics and allow the overall survival rate for both species to increase. He did this by altering the distance between the prey and oranges (their food), establishing barriers to predator movement, and creating corridors for the prey to disperse. These changes increased habitat patches and in turn provided more areas for the prey to seek temporary protection. When the prey would go extinct locally at one habitat patch, they were able to reestablish by migrating to new patches before being attacked by predators. This habitat spatial structure of patches allowed for coexistence between the predator and prey species and promoted a stable population oscillation model. Predator-prey interactions that in turn influence population dynamics have been described in detail

Optimal Foraging Theory

OFT examines the choice of food items. Such prey models (even though the prey doesn't necessarily have to be another animal) look at decisions of basic diet selection. The animal searches for food, and finds potential prey items one at a time. It has to decided whether it should stop searching and eat the food it has found, or whether to ignore the food and keep searching. We assume that the animal is trying to maximise its currency (i.e., rate of energy intake measured in calories) in a particular time period. Each type of food takes a different amount of time to handle it during consumption. For instance, Northwestern Crows search along the water line during low tide for large molluscs called whelks. When it finds one, it flies upward and drops it on the rocks to break it open. Smaller whelks are harder to break, and need to be dropped from a greater hight, so they may be ignored. Also the content of certain kinds of nuts is more difficult to access than that of others.

E_1 and E_2 refer to the caloric energy that is contained within a prey of type 1 or 2 respectively. The handling time needed to access the caloric content for such prey items is h_1 and h_2. If the net energy gained per unit of time (or the rate of energy gain) for type 1 exceeds that for item 2

$$E_1/h_1 > E_2/h_2$$

then you would do well to eat prey of type 1 whenever you find it. If you find an item of type 2 you may consider rejecting it and to keep searching if the energy lost in searching for item 1 added to its handling time is less than what is gained from eating item 2 outright.

$$E_2/h_2 > E_1/(S_1 + h_1)$$

Depending on the relationships of this equation the predator should eat just prey 1 (and be a specialist) or it should eat both prey 1 and prey 2 (and be a generalist). This decision to specialise depends on the quality and abundance of S1, regardless of S2.

Marginal Value Theorem

The MVT considers an optimally foraging creature that exploits patchy resources as it must decide when to move on to the next patch. As it attempts to optimise a cost/benefit ratio individuals will stay longer as the distance between patches increases or when the environment as a whole is less profitable.

Game Theory Models

A more comprehensive prediction of an animal's most profitable behaviour may well require us to also consider what others are doing. A behaviour may well be quite rewarding when it is rare in a population (e.g., deception) but may not be nearly as advantageous to its actor when it becomes common. Game theory models thus consider the value of frequency-dependent fitness benefits. Originally developed as a tool to predict rational human economic behaviour, its application to many evolutionary problems has improved our understanding of situations where fitness consequences of a behaviour depend on types and frequencies of behaviours exhibited by other animals within the population.

It is assumed that individuals represent players of a game in which all parties are aware of the rules, are consciously attempting to maximise their payoffs, and are attempting to predict the moves of their opponents. Players move simultaneously or at least do not observe the other player's move before making their own.

	strategy 1	***strategy 2***
strategy A	3, 3	2, -1
strategy B	-1, 2	2, 2

A Pay-off matrix for a hypothetical, normal form game that pits two individuals against each other. Player Blue (rows) may chose strategy A or B when dealing with animal 2, while player Red (columns) may rely on strategy 1 or 2. The payoff matrix lists the colourcoded consequences that each individual receives for the combination of strategies played.

A Payoff Matrix is used to formally represent a game that includes all conceivable strategies, along with each strategie's corresponding payoff against any other strategy. The matrix of a game specified the payoffs that each player receives for the combinations of actions played. For example, if player 1 plays top and player 2 plays left, player 1 receives 4 and player 2 receives 3. In each cell, the first number represents the payoff to the row player (in this case player 1), and the second number represents the payoff to the column player (in this case player 2).

A case of a symmetric game exists if the payoffs do not depend on which player chooses each action. In the example above, this is the case if strategy A (player 1) is the same as strategy 1 (player 2) and strategy B of player 1 is the same as strategy 2 of player 2. The game can then be represented with a single payoff per cell, i.e., the payoff for the row player. For example, the payoff matrices on the right represents the same game as the matrix above, if 1 = A and 2 = B..

	strategy A	***strategy B***
strategy A	3	2
strategy B	-1	2

The simplified Pay-off matrix rewritten for a hypothetical, symmetrical, normal form game that pits two individuals against each other. As both player may chose strategy A or B, the payoff matrix lists the consequences that the row individual receives for all possible combinations.

Game theory aims to identify behavioural rules that form an Evolutionary stable strategy (or ESS). This strategy, if adopted by a population of players, cannot be invaded by any mutant strategy. An ESS is "evolutionarily" stable meaning that once it is fixed in a population, natural selection alone is sufficient to prevent alternative strategies from invading the population.

Prisoner's Dilemma

Hawks and Doves Game: Game models have greatly increased our understanding of conditions under which ritualised displays may take the place of aggressive interactions. Maynard-Smith's approach involves resource-centered conflict between members of a group which will follow either a strategy of unrestrained use of weapons (hawk) or reliance on highly ritualised elements of fighting (dove):

	Hawk	***Dove***
Hawk	(V"C)/2	V
Dove	0	V/2

Payoff matrix for the Hawk-Dove game. Hawk is a strategy where the individual fights until either it sustains an injury or the opponent retreats. The losing hawks suffers a cost for such injury (C). A Dove contests the interaction with displays but will retreat immediately if the opponent escalates with a use of its weapons. The winner of the encounter obtains the resource in question and gains control over its value (V). Interactions between two hawks result in the term 1/2(V-C) as hawks against hawks will win half of the time and lose half of the time. Moreover, hawks will also sustain an injury half of the time. In interactions with another dove, a dove will win half of the time and retreat without cost the other half of the time.

The different cells of the payoff matrix represent combinations of strategies for the two players, where each would prefer to win, prefer to tie rather than lose, and prefer to lose over injury. Game theory can be used to address the conditions under which such strategies are evolutionarily stable (i.e., an ESS). First we ask if a population playing one strategy can be invaded by a few animals playing the other strategy. To examine whether dove is an ESS we examine whether a population of doves can be invaded by a hawk?

$$u[H,D] > u[D,D]$$

$$V > V/2$$

A hawk can invade a population of doves as its payoff (V) is greater than that of a dove pitted against other doves (V/2). Dove thus cannot be an ESS. Hawks will be able to invade a population of doves until encountering another hawk becomes a common event. To examine whether a dove can invade a population of hawks, we test whether its payoff is greater than that of a hawk when faced with a population of hawks. Formally this evaluates whether

$$u[D,H] > u[H,H]$$

$$0 > (V\text{-}C)/2$$

$$0 > V/2\text{-}C/2$$

$$0 > V\text{-}C$$

The conclusion on whether a dove will be able to invade a population of hawks will thus depend on whether V > C. If the resource value exceeds the costof injury, the benefit to hawks will be positive (i.e., greater than 0), and a population of hawks cannot be invaded. Hawk thus represents an ESS under such conditions. However if V<C then payoff to hawks will be negative and doves can invade. Doves become increasingly good at invading a population of hawks as opportunities for damage increase. In fact a population of hawks would go extinct in situations where the cost of injury exceeds the resource's value. To obtain the proportion of hawk and dove strategies that stably coexist when C>V we examine at what proportion of Hawks (p) and Doves (1-p) the fitness of individuals playing either of the two strategies is equal

$$W[H] = p^*((V\text{-}C)/2) + (1\text{-}p)^*(V) \text{ equals } W[D] = p^*0 + (1\text{-}p)^*(V/2)$$

$$p^*(V/2\text{-}C/2) + (1\text{-}p)^*(V) = (1\text{-}p)^*(V/2)$$

$$p(V/2) - p(C/2) + V - pV = (V/2) - p(V/2)$$

$$p(V/2) + p(V/2) - pV - p(C/2) = (V/2) - V$$

$$- p(C/2) = -(V/2)$$

$$pC = V$$

$$p = V/C$$

Game theory thus provides a theoretical underpinning for why animals tend to only fight with great ferocity when a resource of great value is at stake. Fighting, for instance, is particularly intense in elephant seals where victorious males are able to monopolize a large area of beach which contains many females. In the great majority of instances, however, resources are rarely worth being injured over and competing individuals will resolve most conflict with ritualised displays.

Chicken, Brinkmanship, and War of Attrition Games

	swerve	***straight***
swerve	0	-1
straight	1	-10

The game "Chicken" is a variant of the Hawks and Doves game. In it each player prefers not to yield to the other, but the outcome where neither player yields is by far the worst possible one for both players.

It has its origins in the situation where two drivers head towards each other on a collision course: one must swerve, or both may die in the crash. The driver who does swerve is called a "chicken". The game's most important feature is an assured, and significant cost to the situation where both individuals chose conflict over conciliation. "Losing" by deciding to swerve is so trivial compared to the crash that occurs if nobody swerves, the reasonable strategy would be to swerve. Yet, knowing this one may well decide not to swerve, gambling that the opponent will be reasonable.

Insights gained on the predictability of behaviour based on the presence of overwhelming punitive consequences have formed a central component of the cold war military doctrine of Mutual assured destruction, where the use of nuclear weapons by either side would assure the destruction of the aggressor along with its target. This theory of deterrence states that the deployment of strong weapons as a threat to the enemy is essential in order to prevent the use of the very same weapons. Success in discouraging either side from resorting to these weapons is only effective as it is credible to both sides that a chosing conflict would result in the worst possible outcome for all - nuclear annihilation of both sides.

War of attrition scenarios refer to the situation where fighters attempt to grind down the opponent's defenses. There is no fixed cost associated with losing or contesting, but as the encounter wears on, each player accumulates incremental costs. A decision to give up arises when one indidivual backs down, relinquishing access to the contested resource, rather than continuing to sustain further insults

Pheromones in Ants and Bees

One of the many ways in which organisms can communicate with each other is through the use of pheromones. An organism generates and emits these hormonal chemicals in order to relay a message to another member of the same species. Ants and bees demonstrate two prominent examples of pheromone usage, which acknowledges their incredible capability to organise the behaviours of the whole colony.

Ants produce numerous different pheromones, each with its own distinct purpose. Ants secrete pheromones to attract mates, to signal danger to the colony, or to give directions about a location. Other pheromones act as deterrents keeping out unwanted ants from foreign colonies or preying insectivores. Still other pheromones communicate ants to congregate. This explains how assiduous ants exercise remarkable cooperation in building a colony. Pheromones maintain the cohesiveness and organisation of the complex ant communities. Certain types, such as alarm pheromones, produce a "releaser effect", which induces a quick response and may be used to tell other ants to evacuate a dangerous area such as an approaching lawnmower.

For example, when a spider approaches an ant will release alarm pheromones that alert all the other ants. Ants may also discharge alarm pheromones as a result from being diverted from their work, e.g. heavy human steps. Releaser pheromones are also used to mark territory. As the chemical deposited dries, it signals to other species members of the territory's occupant. Other pheromones create a "primer effect" that entices other ants for actions. Such pheromones are useful in mating rituals and only affect ants of the opposite sex. Primer pheromones can send signals to the endocrine system, to make appropriate changes, for instance ovulation required for successful mating.

Biologist E.O. Wilson discovered in the 1960's that the organic chemical for each pheromone varies tremendously depending on what signal it entails. Ants taste and smell a substance that evaporates off the chemical laid down by another ant. Wilson observed in slow motion films that ants do this by moving their antennae from side to side. For example, alarm pheromones are discharged into the air, and expand in a circle of smell. Ants can determine the concentration of the pheromone, and thus determine the proximity of the source of danger. Bees are also well known for communicating through the use of pheromones. Like ants, bees have a variety of purposes associated with the pheromones, such as marking and behaviour. Honeybees release chemical signals for marking food sources, marking their hive, in scenting potential hive sites, and in assembling swarms for flight.

Although each hive has a particular scent, different colonies can be easily integrated during times when honey flow is heavy since the colony's odor is inundated with the scent of nectar. Virgin queen bees emit a behavioural pheromone released with their feces. When the new queen feels threatened by the workers she uses this pheromone as a repellent. Another widely identified chemical signal in bees is the Queen Mandibular Pheromone (QMP). This pheromone ensures that the queen is the only reproductive female in the hive by compromising the reproductive systems of worker bees. It also provides an attractant signal to the drones.

Lek Polygyny

Lek polygyny is a mating system common in polygynous species of insects and birds in which the male provides no parental care to its offspring. The lek mating system is uniquely driven by the females' pursuit of their mate, rather than the males'. Males of lekking species do not hunt for receptive females. Males form aggregates in neutral locations devoid any resources valuable to females. The group of males performs intricate vocal, visual or chemical displays to lure receptive females to their lekking site. In most lekking species, these group displays typically increase the ratio of visiting females per males. At the lekking site, visiting females are able to compare the males' physiques and courtship displays, picking the most attractive male as their mate (Alcock, 2001). Thus, the few, most attractive males will do the majority of the mating (about 99%), while the subordinate males do no mating at all (Sherman, 1999).

Alternative Hypothesis: Kin Selection Hypothesis

The Lek Polygyny mating system promotes a heavily skewed mating success rate among lekking males. Although most of the individuals in a lek never receive a mating opportunity, lek polygamy continues to flourish among various species of birds and insects (Sherman, 1999). This suggests that the fitness of subordinate males must somehow be indirectly benefited by communal displays. A number of hypotheses have been proposed to explain the reasons behind lekking behaviour. Widely recognised hypotheses include the Hotspot Hypothesis, the Hotshot Hypothesis, and the Female Preference Hypothesis (Alcock, 2001). Petrie *et al.* (1994) proposed an alternative hypothesis predicting that lekking behaviour is driven by kin selection.

In the kin selection hypothesis, Petrie *et al.* (1994) suggests that if all the males in a lek where genetically related, then the males (even the subordinate males) would receive fitness benefits. Petrie *et al.* (1994) demonstrated the kin selection hypothesis by determining the genetic structure of different lekking groups of the Whipsnade Park peacock population. Peacocks, a typical lekking species, form aggregates at neutral display sites. Peacocks use their calls to attract receptive peahens. Upon the arrival of a peahen, the peacocks cease calling and perform intricate plumage displays. As common in lekking species, the peacocks with the most appealing courtship displays have a high mating success while the subordinate peacocks have no mates. The lekking sites of peacocks are carefully chosen and after a male's fourth year of lekking, he will establish his permanent lekking site. Peacocks will return to this same lekking site every mating season (Petrie *et al.*, 1994).

Using multilocus fingerprinting, Petrie *et al.* (1994) compared the genetic similarity of within and between the lekking groups of Whipsnade Park's peacocks. The results supported the proposed kin selection hypothesis. As predicted, the degree of band-sharing within the leks was substantially higher than the band-sharing between the leks. The band sharing within the leks was indicative of that of half-siblings. After this discovery, Petrie *et al.* (1994) proposed subsequent hypotheses of why these lekking groups consisted of related individuals. Petrie el al (1994) suggested that related peacocks tended to display together because their dispersal was concentrated around their natal sites.

This hypothesis was rejected when peacocks of mixed relatedness were reared away from their natal sites but, upon reintroduction into the Whipsnade Park population, joined leks with closely related peacocks. It was then suggested that peacocks' tendency to congregate with relatives was due to a shared genetic preference for habitat-selection. This was rejected due to the homogenous environmental features of the park. Petrie *et al.* (1994) concluded that unique structure of the peacock leks was driven by kin selection based on self-referent phenotypic matching ("Armpit Effect"). Petrie *et al.* (1994) predicted that peacocks match heritable similarities of their own phenotype with those of other males.

Because the authors believed that peacocks do not participate in the rearing of their offspring, they posit that a peafowl's recognition of their father must be genetically innate. The tendency to form leks with relatives, therefore, occurs in the absence of social cues to their identity. By cooperatively forming leks with genetically related individuals, subordinate peacocks forfeit their chances of mating, but increase the chance that their genes will be passed on via successful relatives. The indirect benefits of fitness seen in peacock lekking displays outweigh the costs of communal displays (Petrie *et al.*, 1994).

There are four distinct genera of Peafowl: Pavo, Afropavo, Rheiinartia and Argusianus. Though it is frequentally reported that male peafowl do not participate in nest defense or rearing paternal care of young has been observed in all four genera. Only the common Indian Peafowl Pavo cristatus is frequentally kept in captivity. Its behaviours in zoo settings may not reflect its evolutionary history to the degree that wild populations of the species might.

Hotspot Hypothesis

According to the hotspot hypothesis, males form leks because females frequently visit certain "hotspots". The hotshot hypothesis predicts that males form leks because subordinate males congregate around highly attractive males to increase their chance of being noticed by receptive females.

Female Preference Hypothesis

The female preference hypothesis predicts that males form leks because female like to visit large clusters of males consisting of a variety of potential mates from which she can quickly and safely compare the quality of her mating choices.

Darwin's Finches

The Galapagos Islands finches have been used by Charles Darwin and many other scientist to study how " random variation and natural selection can drive the production of organisms with novel features, adapted to new ways of life (Alberts 2004). The finches studied in the Galapagos Island are usually characterised by the great variance in beaks and the function of beaks. The finches beaks were first thought to change over many years because of local ecological condition, but many scientist devised a new hypothesis which stated, "As a consequence of beak evolution, there have been changes in the structure if finch vocal signals (Podos 2004)." This hypothesis was based the assumption that beaks have a direct correlation in effecting song production.

The Galapagos Island finches displayed a wide variety phenotypes of beaks with different forms and functions caused by natural selection over many years. This phenotype variation was hypothesized by the Grants to be a result of natural selection because of ecological conditions. The new studies, however, tested to see if finch beaks have an affect on the acoustic structure of the songbird songs. The first experiment done to test this hypothesis involved first checking to see if beak movement indeed caused sound production, which in fact turned out to be true. Thus, one could imply that beak variations could cause a different song production. The goal of these researches is " to show how research on the relationship between beaks and song is providing novel insights into the interplay of morphological adaptation and the evolution communication signals (Podos 2004)." The bird song is actually made in the birds syrinx, which is similar to the human larynx during speech production. The scientist then addressed the question of how the birds are able to send pure tonal signals across such a long range and distance.

The answer is that the birds are able to change their vocal tracts configurations, which in turn directly changes the frequency of their songs. The three generalisations, which these finches fall into describing their songs are, simple, variability, and cross-species overlap. If the beak is opened more wide then it creates a higher pitched sound. So it could be hypothesized that a large beaked bird will have a large vocal tract and will in turn create a lower song frequency. This hypothesis was backed up by the studies of Maria Palacious and Pablo Tubaro who tested the Neotropical Woodcreepers Pendrocolaptinae, which produced lower frequency songs with long beaks. These scientists tested, these highly protected finches, by video taping them and then determining the beak gapes for each song for the different finches.

The two main reason male finches use their singing, like many other singing birds, is for territory defense and mate attraction, but the difference in beak size could affect both these characteristics, because some males with either small or large beaks could not produce the right song in other to perform these tasks. This lead Nowick and colleagues to test to see if a beak changed its size for food collection, then would this have a direct affect on how fast or slow a finch could move its beak to produce its specific song. Nowick then concluded that there was a positive correlation between "divergences of beaks and its influence not only the feeding but also their singing behaviour." The scientist, however, plan to broaden their investigation by studying how wider beaks may influence songs and also study other birds with beak morphology such as the crossbills or the Hawaiian honeycreepers.

Territoriality

Territoriality is a type of intraspecific or interspecific competition that results from the behavioural exclusion of others from a specific space that is defended as territory. This well-defined behaviour is exhibited through songs and calls, intimidation behaviour, attack and chase, and marking with scents. This form of defense proves to be very costly for animals. So one is forced to ask, Why do animals take part in such interspecific competition?

In order to understand this question one must take a cost-benefit approach to territoriality. The proximate reasons for such defense vary. For some animals the reason for participating in such elaborate protective behaviour is to acquire and protect food sources, nesting sites, mating areas, or to simply attract a mate.

The ultimate cause of this behaviour may be attributed to the increased probability of survival and reproductive successes. In defending a territory an animal is ultimately securing that it will have an habitat in which to forage for food and to successfully reproduce, thus increasing the animal's overall fitness.

This ultimate theory is strengthened when one considers the instances in which territoriality increases; in times of depleted resources the presence of territoriality increases. The presence of territoriality often forces less fit animals to live in sub optimal habitats, thus reducing their reproductive success.

Though territoriality offers immense reproductive and nutritional benefits, it also comes at a cost. Defending territory is not easy. Territoriality cost time and energy and can often interfere with other fundamental activities as parenting, feeding, courting, and mating. For these reasons territoriality may not be seen as a benefit in all animals. Animals must be able to reap the fruits of territoriality, while expending the least amount of energy. For these reasons if resources are abundant and predictable it would be disadvantageous to defend the territory. On the other hand, if resources are scarce and undependable it would be advantageous to exhibit territoriality.

An animal chooses its territory by deciding what part of its home range it would like to defend. In selecting a territory the size and the quality play a crucial role in determining an animals habitat. Territory size generally tends to be no larger than the organism than requires to survive, because with an increase in territory comes an increased in energy expenditure. For some animals the territory size is not the most important aspect of territoriality, but rather the quality of defended territory. The quality is considered to be fundamentally important due to amount of food availability and superior nesting sights. Animals depend on these features to ensure their superior fitness.

Animals invest a lot of time and energy in defending their territories, and for this reason they fight vigorously to defend their territory at all cost. Researchers suggest for this reason that when a rival challenges a territory holder, the owner almost always wins the contest. This phenomenon could be attributed to an evolutionary stable strategy which assets that rules for behaviour is controlled by an inherited proximate mechanism such that the differences between individuals in their strategies are liked to differences in their genes.

Territory plays an important role as a mechanism of population regulation, insuring the success of fit animals, and aiding in the eradication of less fit animals. Territorially also plays a fundamental role as an indicator of carrying capacity; it also serves as an indicator of how much habitat is necessary to support viable populations. For these reasons researchers continue to examine the well-developed behaviour of territoriality.

Chapter 4

Individuality, Temperament and Genius in Animals

Experimental studies of animal behaviour, pursued for the solution of definite problems of sense, instinct, or habit, frequently yield as by-products interesting and important information concerning individual, sex, species, and race differences. Such observations commonly fail to get recorded because of the primary importance to the observer of the problem on which his attention is focused. In preparing his results for publication he would gladly report everything of significance, were it not that exigencies of time and space render this either impracticable or impossible. It is largely because of our conviction that certain of the unrecorded by-products of our investigations are in various respects more important than the data which we have published, that we are attempting to write on evidences of individuality in various organisms.

In this field of naturalistic and experimental observation, there is always the risk of confusing age, sex, or race differences with those which are truly individual. The casual observer readily overlooks the fact that his pet canaries, kittens, or dogs, differ by several weeks in age or are otherwise not suitable for comparison, for as a naturalist he is less concerned with strict comparability than with that knowledge which will lead to sympathetic insight. But to those who are trained in critical and well-controlled observation, it is an easy task to eliminate such sources of error and to obtain fairly comparable data concerning individuality. Field naturalists and the born lovers of animals know by intimate acquaintance that important individual differences exist in many species of organism, but experimentalists are less generally aware of this fact, for their attention tends to be monopolised by problems of species characteristics and of general organic functions or reactive capacity.

Field naturalists and the born lovers of animals know by intimate acquaintance that important individual differences exist in many species of organism, but experimentalists are less generally aware of this fact, for their attention tends to be monopolised by problems of species characteristics and of general organic functions or reactive capacity.

Even in invertebrates individuality becomes conspicuous with familiarity. Among earthworms we have observed that specimens, comparable in all essential points and existing under the same conditions of observation, exhibit surprisingly different modes of response. Thus, one individual adapts itself to the demands of a situation, works smoothly, steadily—as it were willingly; another slowly and haltingly meets the experimenter's requirements. Its tendency to do the wrong thing seemingly amounts to perversity or stubbornness. And so the observer gains the feeling that the two organisms are quite as different in reactive tendency as are two men.

It has often been remarked that the individuals of a human race with which one is unfamiliar look alike. This we always discover to be due to our failure to notice marked individual differences. As our familiarity with the type increases, these individual traits become increasingly obvious. Now precisely what is true in our experience with our fellow men is still more true of other types of organism. We note at first only the species or racial differences, or perhaps if they be equally conspicuous, certain age and sex differences, but as we continue to live with the organisms and to observe them carefully day by day, we come to appreciate those qualitative and quantitative peculiarities which constitute individuality. As far as we can see, there is no significant difference in degree of individuality between earthworm and man, ant and monkey.

Even in invertebrates individuality becomes conspicuous with familiarity. Among earthworms we have observed that specimens, comparable in all essential points and existing under the same conditions of observation, exhibit surprisingly different modes of response.

Intimacy of relation with a wide range of organic types has served, among other things, to convince us that temperament, character, and genius are terms, which, like individuality, may be used quite as appropriately in connection with one type of organism as with another. We wish especially, in this paper, to report certain of our observations concerning these aspects of life. Temperament we have come to think of as the sum of fundamental, inborn reactive tendencies,—they are sometimes called primary instincts; character, as these same tendencies organised through environmental contact or experience into a complex and more or less highly adaptive system of behaviour; genius, as exceptionally strong or well-marked temperamental traits of a particular order. The conventional and ancient classification of temperaments according to strength and duration of response as choleric, melancholic, sanguine, and phlegmatic seems unduly simple in the light of our observations, for there are at least several important ingredients or constituents of temperament which apparently vary independently or in groups with respect to strength and duration of response, and possibly also in other important ways. We may not here further dwell upon definitions, but we shall hope to render these suggestions more significant by the facts which we have to record.

Some years ago we undertook a comparative study of two strains of albino rat, the one closely inbred for many generations, the other outbred. Save for this difference, the individuals of the strains were entirely comparable. We attempted by various experimental means to discover peculiarities of behaviour in these animals. Soon it became apparent that the inbred individuals adapted themselves less readily to new environmental demands. They proved less apt pupils in tests of habit formation. We were struck, as our observations progressed, by certain peculiarities of behaviour which appeared to be characteristic of the strains rather than of individuals.

Among them fear, timidity, savageness, curiosity, sociability were conspicuous. In general, the inbred rats seemed more timid, fearful, more likely to defend themselves by biting if disturbed, One rat is extremely fearful of anything new or unusual, it shrinks timidly from the experimenter. It can only with difficulty be induced to run its way through the experimental apparatus. Another individual of the same age and sex, born in the same litter, is by contrast aggressive and exhibits marked initiative in new situations. less ready to try new things, more suspicious of the experimenter, slower to acquire obviously profitable modes of response than were the outbred animals. These differences in behaviour seemed to us to account for an apparent difference in intelligence, and we finally concluded that it is really quite beside the mark to contrast the two strains by saying that the one is the more or the less intelligent.

Subsequently, increasingly definite and well-controlled studies were made, in which were recorded observations concerning the preferred position of an individual in its cage or nest box; the relative positions at different hours of the two individuals, male and female, in a given cage; the degree or amount of activity; savageness, or the tendency to bite; and the quickness and amount of response to various stimuli. These and similar observations shortly indicated that savageness designates certain tendencies to reaction, as does also fear, timidity, or wildness, and that our only intelligible way of defining these terms is by enumerating the several types of activity.

Wildness, for example, is indicated by attempts to hide in the cage or in the observer's hand, random and excited running about with repeated attempts to escape, squeaking, and various other forms of response. Timidity, which seemingly is not identical with wildness or fear, involves the avoidance of the experimenter, a kind of chattering or gnashing of the teeth, cowering, or even trembling.

Although most of our studies have been concerned with relations of behaviour to inbreeding or to the crossing of individuals which differ markedly in some trait, we have incidentally obtained abundant evidence of important individual differences of the temperamental sort. One rat for example, is extremely fearful of anything new or unusual, it shrinks timidly from the experimenter. It can only with difficulty be induced to run its way through the experimental apparatus. When cornered, it defends itself by biting the experimenter's hand. Its wildness is indicated by persistent efforts to hide or to escape.

It responds quickly and markedly to any sudden and unexpected stimulus; peculiarly startling stimuli at times cause it to tremble. Another individual of the same age and sex, born in the same litter, is by contrast aggressive and exhibits marked initiative in new situations. Its fear or timidity is readily overcome by its curiosity. It quickly becomes accustomed to the experimenter, and allows him to touch it or take it up in his hand without attempting to bite, and shortly without effort to escape. It responds slowly and only slightly to most stimuli and is disturbed only by strong stimulation. In a word, the two rats are temperamentally as different as any two human beings one is likely to meet. It is such observations as these, made on many different individuals, that have wholly convinced us of the desirability of a careful analysis of temperament and the reduction to terms of measured description of its chief constituents.

We once undertook to study experimentally the ideational behaviour of pigs. For this purpose two young animals were chosen, the one a male, the other a female. They were observed daily, and for several hours each day, the whole of one summer. We became sufficiently well acquainted with their characteristics to appreciate alike their varying degrees of intelligence and their temperamental peculiarities. What we have not published in our report on the behaviour of these creatures certainly would interest the general reader much more than our printed contribution toward the solution of our problem. We therefore venture to present certain of the fascinating byproducts of our summer's work. That the differences which we are about to emphasize are not necessarily individual, we readily admit; that they are not age or species differences, we are certain. We expect that some, at least, are sex differences.

Nip and Tuck, for thus we early decided to designate our subjects, soon made us feel their individuality. Both, under the spur of the hunger motive, worked remarkably well toward the solution of ideational problems, and their success in this work fully justified the popular impression that the pig is one of the more intelligent among mammals. Nip, the male, was less active and energetic than his sister, Tuck. He also was less greedy, and showed rather less initiative and a more limited range of reactions. Tuck it was who usually led if there was opportunity for competition, while Nip followed.

Both quickly became accustomed to the experiment, but Nip showed more persistent wariness, timidity, and suspicion than did Tuck. She, however, was much quicker in response, more alert, curious, and quick to discover new opportunities for pig satisfaction. When at work on experimental problems, Nip much more easily discouraged and tended earlier than Tuck to give up his search for the reward of success. Tuck constantly acquired new and profitable tricks, which as a rule sooner or later appeared in Nip also, sometimes spontaneously and again by reason of the imitative tendency.

One summer we removed a brood of four young crows from their nest just before they were able to fly. We could not identify the sexes at the time, so the differences we observed may be either sex or individual, but at any rate, the four specimens were as sharply contrasted in temperament as are the children of any household.

As day after lay we observed these two specimens of a mammalian type whose life under domestication gives its intelligence slight opportunity for display, we were strongly impressed, as we had been in the case of rats also, by the importance of temperamental reactive tendencies in responses to any experimentally arranged situation. The experimenter who ignores individuality or temperament in his subjects runs a grave risk of misunderstanding or wrongly evaluating his results. Our descriptions sound anthropomorphic, but that, the alert reader will appreciate, is due to our avoidance of stilted and unnatural terminology. We are attempting to describe in an intelligible way, and briefly, behaviour which, if we should restrict ourselves to wholly objective terms, would require pages of unusual behaviouristic statement.

Among the birds, there is probably no more interesting object of study than the crow. Its species characteristics are notably alluring to the psychologist, but to us, as a result of varied observations in the corn fields of Pennsylvania and on the wooded hills of New Hampshire, sex, age, and individual differences are no less fascinating. One summer we removed a brood of four young crows from their nest just before they were able to fly. We could not identify the sexes at the time, so the differences we observed may be either sex or individual, but at any rate, the four specimens were as sharply contrasted in temperament as are the children of any household.

We set about rearing these birds by hand, the while taming them for experimental purposes. Within a few days, one of the four began to exhibit the characteristic fear reaction of its species, and at once it became extremely difficult to feed. For a few days we persisted in our attempts, and then as he or she, as the case may have been, was no less persistent, we decided to devote our time and energies to his three companions. Thus, at the very outset, temperamental peculiarities, perhaps amounting to nothing more than exceptionally strong and persistent fear reactions, served to eliminate one of the individuals from our collection.

Our space will not permit us to recite in detail, as we are tempted to do, the peculiarities which these birds exhibited during a memorable summer. We must content ourselves with the simple statement that in reactions which may be designated as those of wildness, fear, timidity, curiosity, suspicion, initiative, sociability, the individuals differed most obviously and importantly. We hope sometime, in justice to the problem of crow temperament, to devote a summer to the intensive study of sex and individual differences in these extremely intelligent birds.

Concerning temperament, character, and genius in the Primates, our materials are at once abundant and to us absorbing. Everyone who knows anything about primates, high or low, realizes that in them individuality is more conspicuous for the human observer than in most other organisms. But our results do not justify the conclusion that temperamental differences are more obvious or more important in monkeys, anthropoid apes, or man, than in crows, pigs, or rats. We have come to suspect that the popular opinion concerning the matter is due chiefly to similarity of structure and behaviour—in a word, to felt kinship. It is simply because we are more like monkeys and apes that we more readily notice and more highly value their individual characteristics.

Not so very long ago, we had a splendid opportunity to become intimately acquainted with two adult male monkeys of the species *Pithecus irus [Macaca irus]*. The one, we shall call Skirrl; the other, Jimmie. It would be easier to tell what these individuals had in common than to enumerate their differences. Their temperamental divergences constantly amazed us. But here we must content ourselves with an account of a few of the most remarkable differences in behaviour.

The monkey was provided with a suitable hammer, nails, and a board. He went to work immediately and although be exhibited no constructive ability, his skill, without tuition, in handling hammer and nails and in driving the latter into the board or elsewhere, according to his taste, more than equaled that of the unpracticed human.

Skirrl's attitude toward the friendly experimenter was frankly aggressive, but not vicious. Jimmie was extremely vicious; he never could be trusted. Skirrl's interest in objects which he could play with or in any wise manipulate proved inexhaustible, whereas Jimmie exhibited slight interest in other objects than the members of his species, his enemies, or foods. By a competent observer who had studied him carefully prior to our acquaintance, we were told that Skirrl was feeble-minded. And it certainly seemed so, when, as frequently happened, he sat before an experiment box, yawning repeatedly, and from time to time interrupting these expressions of ennui by half-hearted attempts to solve his problem. Whereas Skirrl rather quickly became accustomed to unusual experimental situations, Jimmie was so wary and distrustful that we finally gave up our attempts to observe his behaviour under rigidly controlled conditions, and treated him merely as a visitor in the laboratory.

One day we noticed Skirrl pounding with a stick a nail which he had found in his cage. We were quick to follow this cue. The monkey was provided with a suitable hammer, nails, and a board. He went to work immediately and although be exhibited no constructive ability, his skill, without tuition, in handling hammer and nails and in driving the latter into the board or elsewhere, according to his taste, more than equaled that of the unpracticed human. In the presence of the same outfit of tools, Jimmie threw the hammer into one corner of his cage, scattered the nails about the floor, and proceeded to tear the board to pieces with his teeth. Never did he exhibit the least inclination to use hammer and nails independently or together as tools or implements.

When given a saw, rendered indestructible by metallic guards for the handle, and a heavy wooden block on which the saw might be used, Skirrl was manifestly pleased. He used the saw in quite as many and varied ways as might a boy of four or five years. By sawing before him at various times, the observer tried to teach him to use it in the conventional human way. But to this he paid scant attention, preferring, it seemed, to work out his own modes of amusement. Finally, he hit upon a way of using the saw which we have been told is in vogue among certain peoples of the East. Sitting on the floor, he held it teeth uppermost, his feet grasping the handle tightly and holding the saw firmly in position. He then grasped a nail by both ends and rubbed it rapidly over the teeth of the saw, thus producing a noise which evidently delighted his soul.

It is clear enough from the responses of other monkeys of the same and opposite sex (the same and other species) to saw, hammer, and nails, as well as to other implements, that Skirrl's behaviour must be described as highly individual or temperamental. Never have we observed anything comparable with it in any untaught Primate other than the human. We have agreed to call Skirrl's behaviour an expression of genius, for the more we consider the matter the more certain we feel that this particular individual possesses remarkably strong tendencies to react to certain objects as tools or mechanical devices. From our point of view, he possesses an unusual type of interest or the same to an unusual degree. Feeble-minded though he may be as far as most intellectual requirements are concerned, he is a genius in mechanical manipulation, and to him we feel indebted for a new point of view and for new insight into the meaning of genius.

The anthropoid apes are so manlike in appearance and behaviour that we should be surprised were they not highly individualised and possessed of temperamental traits as well as forms of genius strikingly similar to our own. Our opportunities for intimate acquaintance with the higher apes have been disappointingly few, but with one young orang-utan whom we knew as Julius, we came into delightfully friendly relations. Julius was not born in captivity—few anthropoid apes have that advantage, or disadvantage—but he was captured young, and when we knew him in California, he was probably not far from five years old. Already we have recorded in print many interesting features of his behaviour, as well as our strong conviction of the supreme importance to science and to other aspects of civilization of the thorough study of the anthropoid apes.

When threatened with punishment or actually punished, and when, out of sorts or ill, the young orang-utan behaved so like a child of two or three years that he caused the observers to feel uncomfortably sympathetic.

Julius one day was resting placidly in his good-size cage. A workman passing the cage stopped and offered him a banana. He hurried over to get the proffered food, but just as he reached out his hand for it, the man unkindly drew it away and started to walk off. Julius, evidently disappointed and seemingly resentful, turned, and by a series of somersaults rapidly rolled the whole length of his cage. Later, the same sort of behaviour was observed in quite different situations. When, for instance, after working persistently to solve an experimental problem, he failed to obtain the desired reward of food, Julius would bring his head to the floor with a thump and turn a few somersaults. In thus expressing his feelings of disappointment and resentment, he seemed to relieve himself, for afterward be would go to work, sometimes with energy and a fair show of cheerfulness. It may be remarked, by the way, that similar modes of response have been observed in children of two to six years of age. We recall an instance in which a little boy who for some time had been working unsuccessfully on an ideational problem bumped his head several times, carefully it is true, against a wooden partition, and then remarked, by way of explanation, that he wished to stir things up.

When threatened with punishment or actually punished, and when, out of sorts or ill, the young orang-utan behaved so like a child of two or three years that he caused the observers to feel uncomfortably sympathetic. Many aspects of his behaviour, which unhappily we may not now stop to describe, reminded us of our observations of children, and we found ourselves involuntarily comparing him with human subjects.

How surprising it is, as one stops to reflect on this matter of temperament, that in the same household, as children of the same parents, we find individuals who seem to be opposites in the most varied respects. The one child is sympathetic; the other tends to be cold, unresponsive, or even cruel. The one is frank, naturally honest; the other sly, secretive, and unreliable. The one kindly, good-natured; the other sour and resentful. As these children develop, their temperamental traits may be molded perhaps by educational influences into equally valuable types of character. But never by any chance can they come to possess similar temperamental characteristics.

Surely we shall do well to observe diligently and develop means of studying carefully and measuring the various constituents of temperament, and the factors which enter into character. We should study the constitution and varieties of genius, and especially the conditions which, as experience, operate upon temperamental traits to develop the responses of genius and to elaborate character. For in our efforts to control and direct human life knowledge of those aspects of individuality is of fundamental importance, and there are today unmistakable indications that the future will require of us a science of human behaviour which shall consider as carefully the individual as the species. We live in the era of the biological sciences, and we look forward to an unprecedented development of the sciences of organic function, and especially of those which, like physiology, psychology, and sociology, attempt to inform us concerning phenomena of behaviour. These sciences promise to become of supreme importance to civilization.

Neutering

Neutering, from the Latin *neuter* (of neither sex), is the removal of an animal's reproductive organ, either all of it or a considerably large part. The process is often used in reference to males whereas *spaying* is often reserved for females. Colloquially, both terms are often referred to as *fixing*. While technically called *castration* for males, in male horses, the process is referred to as *gelding*.

Neutering is the most common sterilizing method in animals. In the United States, most humane societies, animal shelters and rescue groups urge pet owners to have their pets spayed or neutered to prevent the births of unwanted litters, contributing to the overpopulation of unwanted animals in the rescue system.

Health and Behavioural Effects

Advantages: Besides being a birth control method, and being convenient to many owners, neutering/spaying has the following health benefits:

- Sexually dimorphic behaviours such as mounting, urine spraying and some forms of male aggression (relating to females in estrus) may be reduced due to the decrease in hormone levels brought about by neutering. This is especially significant in male cats due to the extreme undesirability of male cat sexual behaviour for many pet owners.
- Prevention of mammary tumors: Female cats and dogs are about seven times less likely to develop mammary tumors if they are spayed before their first heat cycle. Mammary neoplasia, or breast cancer, is a very common disorder of female dogs, with a reported incidence of 3.4%. Of female dogs with mammary tumors, 50.9% have malignant tumors. Female dogs that have been spayed before their first heat have a lifetime chance of developing mammary tumors of about 99.5% less than that of intact females. If allowed to go through their first heat before

spaying, then their risk is close to 92% less. Also, spaying female dogs more than two years before the removal of mammary tumors increases the dog's survival odds by 45%.

- Without its ability to reproduce, a female animal effectively has a zero risk of pregnancy complications, such as spotting and false pregnancies, the latter of which can occur in more than 50% of unspayed female dogs.

Disadvantages

General:

- As with any surgical procedure, immediate complications of neutering include the usual anesthetic and surgical complications, such as bleeding and infection. These risks are relatively low in routine spaying and neutering; however, they may be increased for some animals due to other pre-existing health factors. In one study the risk of anesthetic-related death (not limited to neutering procedures) was estimated at 0.05% for healthy dogs and 0.11% for healthy cats. The risk for sick dogs and cats were 1.33% and 1.40% respectively.
- Spaying and neutering dogs and cats increases the risk of obesity. In cats, a decrease in sex hormone levels seems to be associated with an increase in food intake. In dogs, the effects of neutering as a risk factor for obesity vary between breeds.
- Neutered dogs of both sexes are at a twofold excess risk to develop osteosarcoma as compared to intact dogs.
- Studies of cardiac tumors in dogs showed that there was a 5 times greater risk of hemangiosarcoma, one of the three most common cancers in dogs, in spayed females than intact females and a 2.4 times greater risk of hemangiosarcoma in neutered dogs as compared to intact males.
- Spaying and neutering is associated with an increase in urinary tract cancers in dogs.
- Neutered dogs of both sexes have a 27% to 38% increased risk of adverse reactions to vaccinations. The incidence of adverse reactions for neutered and intact dogs combined is 0.32%.
- Neutered dogs have also been known to develop hormone-responsive alopecia (hair loss).
- A 2004 study found that spayed and neutered dogs had a higher incidence of CCL rupture, a form of ACL injury.

Specific to Males:

- About 2% of neutered male dogs eventually develop prostate cancer, compared to less than 0.6% of intact males. The evidence is most conclusive for Bouviers.
- In a study of 29 intact male dogs and 47 castrated males aged 11–14, the neutered males were significantly more likely to progress from one geriatric cognitive impairment condition (out of the four conditions – disorientation in the house or outdoors, changes in social interactions with human family members, loss of house training, and changes in the sleep-wake cycle) to two or more conditions. Testosterone in intact males is thought to slow the progression of cognitive impairment, at least in dogs that already have mild impairment.
- As compared to intact males, male neutered cats are at an increased risk for certain problems associated with feline lower urinary tract disease, including the presence of stones or a plug in the urethra and urethral blockage.
- Neutering also has been associated with an increased likelihood of urethral sphincter incontinence in males.

Specific to Females:

- Spayed female dogs can develop urinary incontinence. Studies report incidence rates of 4.9% to 20%.
- Spayed female dogs are at an increased risk of hypothyroidism.

Current Research

Various studies of the effects neutering has overall on male and female dog aggression have been unable to arrive at a consensus. A possible reason for this according to one study is changes to other factors have more of an effect than neutering. One study reported results of aggression towards familiar and strange people and other dogs reduced between 10 and 60 percent of cases, while other studies reported increases in possessive aggression and aggression towards familiar and strange people, and yet another study reported no effect on territorial aggression, and only a reduction in dominance aggression that existed for at least 5 years. For females with existing aggression, many studies reported increases in aggressive behaviour and some found increased separation anxiety behaviour. A report from the American Kennel Club Canine Health Foundation reported significantly more behavioural problems in castrated dogs. The most commonly observed behavioural problem in spayed females was fearful behaviour and the most common problem in males was aggression. Early age gonadectomy is associated with an increased incidence of noise phobias and undesirable sexual behaviours.

Methods

Females (Spaying): In female animals, spaying involves abdominal surgery to remove the ovaries and uterus (hystero-oophorectomy). Alternatively, it is also possible to remove only the ovaries (oophorectomy), which is mainly done in cats and young female dogs. Spaying is performed commonly on household pets such as cats and dogs, as a method of birth control. It is performed less commonly on livestock, as a method of birth control or for other reasons. In mares, these other reasons include behaviour modification. The surgery is usually performed through a ventral midline incision below the umbilicus. The incision size varies depending upon the surgeon and the size of the animal. The uterine horns are identified and the ovaries are found by following the horns to their ends.

There is a ligament that attaches the ovaries to the 13th rib which may need to be broken so the ovaries can be identified. The ovarian arteries are then ligated with resorbable suture material and then the arteries transected. The uterine body (which is very short in litter bearing species) and related arteries are also tied off just in front of the cervix (leaving the cervix as a natural barrier). The entire uterus and ovaries are then removed. The abdomen is checked for bleeding and then closed with a 3 layer closure. The linea alba and then the subcutaneous layer are closed with resorbable suture material. The skin is then stapled, sutured, or glued closed. Spaying in female dogs removes the production of progesterone, which is a natural calming hormone and a serotonin uplifter. Spaying may therefore escalate any observable aggressive behaviour, either to humans or other dogs. The risk of infections, bleeding, ruptures, inflammation and even reactions to the drugs given to the animal as part of the procedure are all possibilities that should be considered.

Males (Castration)

In male animals, castration involves the removal of the testes, and is commonly practiced on both household pets (for birth control and behaviour modification) and on livestock (for birth control, as well as to improve commercial value).

Nonsurgical Alternatives

Injectable:

- Male dogs – Neutersol (Zinc gluconate neutralised by arginine). Cytotoxic; produces infertility by chemical disruption of the testicle. It is now produced as Esterilsol in Mexico.

- Male rats – Adjudin (analogue of indazole-carboxylic acid), induces reversible germ cell loss from the seminiferous epithelium by disrupting cell adhesion function between nurse cells and immature sperm cells, preventing maturation.
- Male sheep and pigs – Wireless Microvalve. Using a piezoelectric polymer that will deform when exposed to a specific electric field broadcast from a key fob (like a car alarm) the valve will open or close, preventing the passage of sperm, but not seminal fluid. Located in a section of the vas deferens that occurs just after the epididymis, the implantation can be carried out by use of a hypodermic needle.
- Female mammals – Vaccine of antigens (derived from purified Porcine zona pellucida) encapsulated in liposomes (cholesterol and lecithin) with an adjuvant, latest US patent RE37,224 (as of 2006-06-06), CA patent 2137263 (issued 1999-06-15). Product commercially known as SpayVac, a single injection causes a treated female mammal to produce antibodies that bind to ZP3 on the surface of her ovum, blocking sperm from fertilizing it for periods from 22 months up to 7 years (depending on the animal). This will not prevent the animal from going into heat (ovulating) and other than birth control, none of the above mentioned advantages or disadvantages apply.

Other

- Male mammals - Noninvasive vasectomy using ultrasound.
- Female mammals - orally administered phosphodiesterase 3 inhibitor ORG 9935 daily before and during ovulation which blocks the resumption of meiosis resulting in ovulation of a non-fertilizable, immature oocyte without rupturing the follicle.

Surgical Alternatives

Vasectomy: The cutting and tying of the vasa deferentia. Failure rates are insignificantly small. This procedure is routinely carried out on male ferrets and sheep to manipulate the estrus cycles of in-contact females. It is uncommon in other animal species.

Tubal Ligation: Snipping and tying of fallopian tubes as a sterilisation measure can be performed on female cats and dogs. Risk of unwanted pregnancies is insignificantly small. Only a few veterinarians will perform the procedure.

Like other forms of neutering, vasectomy and tubal ligation eliminate the ability to produce offspring. They differ from neutering in that they leave the animal's levels and patterns of sex hormone unchanged. Both sexes will retain their normal reproductive behaviour, and other than birth control, none of the advantages and disadvantages listed above apply. This method is favored by some people who seek minimal infringement on the natural state of companion animals to achieve the desired reduction of unwanted births of cats and dogs.

Penile translocation is sometimes performed in cattle to produce a "teaser bull", which retains its full libido, but is incapable of intromission. This is done to identify estrous cows without the risk of transmitting venereal diseases.

Early-age Neutering

Early-age neutering (or prepubertal gonadectomy – the removal of the ovaries or testes before the onset of puberty) is typically performed in dogs and cats between 8 and 16 weeks of age, as compared to the conventional 6 to 8 months. It is used mainly in animal sheltering and rescue where puppies and kittens can be neutered before being adopted out, eliminating non-compliance with sterilisation agreement, which is typically above 40%. The American Veterinary Medical Association, American Animal Hospital Association and the Canadian Veterinary Medical Association support the procedure for population control, provided that the veterinarian uses his/her best knowledge when making the decision about the age at neutering.

While the age-unrelated risks and benefits cited above also apply to early-age neutering, various studies have indicated that the procedure is safe and not associated with increased mortality or serious health and behavioural problems when compared to conventional age neutering. Anesthesia recovery in young animals is usually more rapid and there are fewer complications. One study found that in female dogs there is an increasing risk of urinary incontinence the earlier the procedure is carried out; the study recommended that female dogs be spayed no earlier than 3 to 4 months of age. A later study comparing female dogs spayed between 4 to 6 months and after 6 months showed no increased risk.

One study showed the incidence of hip dysplasia increased to 6.7% for dogs neutered before 5.5 months compared to 4.7% for dogs neutered after 5.5 months, although the cases associated with early age neutering seems to be of a less severe form. There was no association between age of neutering and arthritis or long-bone fractures. Another study showed no correlation between age of neutering and musculoskeletal problems. A study of large breed dogs with cranial cruciate ligament rupture associated early-age neutering with the development of an excessive tibial plateau angle. Female dogs neutered early are much more likely to develop cystitis although the risk does not appear to be chronic. Two studies showed an increased risk of canine parvovirus infection, which one of the study attributed to the increased susceptibility of young dogs rather than long term immune suppression.

In terms of behaviour in dogs, separation anxiety, aggression, escape behaviour and inappropriate elimination are reduced while noise phobia and sexual behaviour was increased. In males with aggression issues, earlier neutering may increase barking. In cats, asthma, gingivitis, and hyperactivity were decreased, while shyness was increased. In male cats, occurrence of abscesses, aggression toward veterinarians, sexual behaviours, and urine spraying was decreased, while hiding was increased.

Terminology for Neutered Animals

Male Animals: Neutered males of given animal species sometimes have specific names:

- Dog
- Cat, ferret – Gib
- Cattle – Bullock, Ox, Steer, Stag
- Chicken – Capon
- Deer – Havier
- Goat – Wether
- Horse, Donkey – Gelding
- Pig – Barrow
- Rabbit – Lapin
- Sheep – Wether.

Female Animals

A specialised vocabulary in animal husbandry and fancy has arisen for spayed females of given animal species:

- Chicken – Poulard
- Ferret – Sprite.

Religious Views on Neutering

Islam

While there are differing views in Islam with regard to neutering animals, some Islamic associations have stated that when done to maintain the health and welfare of both the animals and the community, neutering is allowed on the basis of 'maslaha' (general good) or "choos[ing] the lesser of two evils".

Judaism

Traditional interpretations of Orthodox Judaism forbids the castration of both humans and non-human animals by Jews, except in lifesaving situations. In 2007, the Sephardic Chief Rabbi of Israel Rabbi Shlomo Amar issued a ruling stating that it is permissible to have companion animals spayed or neutered on the basis of the Jewish mandate to prevent cruelty to animals.

Frog Hearing and Communication

Frogs and toads produce a rich variety of sounds, calls, and songs during their courtship and mating rituals. The callers, usually males, make stereotyped sounds in order to advertise their location, their mating readiness and their willingness to defend their territory; listeners respond to the calls by return calling, by approach, and by going silent. These responses have been shown to be important for species recognition, mate assessment, and localisation. Beginning with the pioneering experiments of Robert Capranica in the 1960s using playback techniques with normal and synthetic calls, behavioural biologists and neurobiologists have teamed up to use frogs and toads as a model system for understanding the auditory function and evolution. It is now considered an important example of the neural basis of animal behaviour, because of the simplicity of the sounds, the relative ease with which neurophysiological recordings can be made from the auditory nerve, and the reliability of localisation behaviour. Acoustic communication is essential for the frog's survival in both territorial defense and in localisation and attraction of mates. Sounds from frogs travel through the air, through water, and through the substrate. The neural basis of communication and audition gives insights into the science of sound applied to human communication.

Sound Communication

Behavioural Ecology: Frogs are more often heard than seen, and other frogs (and researchers) rely on their calls to identify them. Depending on the region that the frog lives in, certain times of the year are better for breeding than others, and frogs may live away from the best breeding grounds when it is not the species' mating season. During the breeding season, they congregate to the best breeding site and compete for call time and recognition. Species that have a narrow mating season due to ponds that dry up have the most vigorous calls.

Calling Strategy

Male-male Competition: In many frog species only males call. Each species has a distinct call, though even among the same species, different dialects are found in different regions. Although humans cannot detect the differences in dialects, frog distinguish between regional dialects. For example, male bullfrogs can recognise the calls of their direct territorial neighbors. By ignoring the calls of these neighbors they save energy, and only vocalize aggressively with to an intruder's call. In this way, calls establish territories, but they also attract females. Males may have a solitary call for times when there is no competition that uses less energy. During other times, when a frog must compete with hundreds or thousands of other frogs to be heard, together they perform a chorus call where each frog calls in turn, successively. The most important feature of the chorus is the shared pattern. Through this pattern, few individuals calls are drowned out. One frog's call may be dominant and trigger the calls of the responding frogs in symphony. Interestingly, calling is linked to physical size and females may be attracted to more vigorous calls. Frogs in the same region chorus within their species and between different species. Frogs of the same species will retune their frequency so it is distinct from other frogs of the same species. Different species of frogs living in the same region have more dramatically different call frequencies. The frequency and durations of different species' calls vary similarly to the preference of that species' females. The neural circuity of females of different species varies.

Male-female Interactions

Like the males, females can distinguish the minute differences between individual frogs. However, males and females are attuned to different parts of the advertisement call. For example, males of the onomatopoeically named species *Coqui* are more attuned to the low frequency *co* part of the call, whereas females are more attuned to the high frequency *qui*. In fact, the order of the parts does not matter. Similarly, for females of the *Tungara* species, the female basilar papilla is biased towards a lower-than-average "chuck" portion of a male call. Experiments that measure the vocal responses and approaches shows these attenuations.

Mode of Sound Communication

Calls are often sent through the air, but other mediums have been discovered. Some species call while they are under water and the sound travels through the water. This is adaptive in a region with many species competing for air time. Narins has found female frog species that use solid surfaces, such as blades of grass and logs, upon which they tap rhythmically to attract mates. Also, Feng has found that some species of frogs use.

Sound Production

The smallest frogs must consume lots of energy to produce calls. In addition, vocalizing muscles can make up 15% of a male spring peeper's body mass, while the same muscles are only 3% of females. Frogs produce sound from the air sac below their mouth that from the outside, is seen to inflate and deflate. Air from the lungs is channeled to the air sac, which resonates to make the sound louder. The larynx is larger and more developed in males, though not significantly different from females.

Sound Localisation

Biologists believed that frogs ears are placed too close together to localise sound accurately. Frogs cannot hear short, high frequency sounds. Sound is localised by the time difference when the sound reaches each ear. The "vibration spot" near the lungs vibrates in response to sound, and may be used as an additional measure to localise from.

Applications of frog Neuroethology

Dr. Feng's work applies the neuroethology of frog communication to medicine. A recent project on hearing aids is based on how female frogs find their mates. Females must recognise the male they choose by his call. By localising where his call is coming from she can find him. An additional challenge is that she is localising his call while listening to the many other frogs in the chorus, and to the noise of the stream and insects. The breeding pond is a very noisy place, and females must distinguish a male's calls from the other noise. How they recognise the sound pattern of the male they are pursuing from the surrounding noise is similar to how intelligent hearing aids help people hear certain sounds and cancel out others. The underlying neural mechanisms are fast neural oscillations, and synaptic inhibition to cancel out noise. The timing and frequency of the sound also play a part in frog communication and may be used in Feng's work. He also studies bat echolocation to create intelligent hearing aids. He is also working on cochlear implants.

Animal Echolocation

Echolocation, also called biosonar, is the biological sonar used by several kinds of animals. Echolocating animals emit calls out to the environment and listen to the echoes of those calls that return from various objects near them. They use these echoes to locate and identify the objects. Echolocation is used for navigation and for foraging (or hunting) in various environments.

Echolocating animals include some mammals and a few birds; most notably microchiropteran bats and odontocetes (toothed whales and dolphins), but also in simpler form in other groups such as shrews, one genus of megachiropteran bats (*Rousettus*) and two cave dwelling bird groups, the so called cave swiftlets in the genus *Aerodramus* (formerly *Collocalia*) and the unrelated Oilbird *Steatornis caripensis*.

The term *echolocation* was coined by Donald Griffin, whose work with Robert Galambos was the first to conclusively demonstrate its existence in bats in 1938. Long before that, however, the 18th century Italian scientist Lazzaro Spallanzani had, by means of a series of elaborate experiments, concluded that bats navigate by hearing and not by vision. Echolocation in odontocetes was not properly described before two decades later, by Schevill and McBride.

Basic Principle

Echolocation is the same as active sonar, using sounds made by the animal itself. Ranging is done by measuring the time delay between the animal's own sound emission and any echoes that return from the environment. The relative intensity of sound received at each ear as well as the time delay between arrival at the two ears provide information about the horizontal angle (azimuth) from which the reflected sound waves arrive. Unlike some man-made sonars that rely on many extremely narrow beams and many receivers to localise a target (multibeam sonar), animal echolocation has only one transmitter and two receivers (the ears). Echolocating animals have two ears positioned slightly apart. The echoes returning to the two ears arrive at different times and at different loudness levels, depending on the position of the object generating the echoes. The time and loudness differences are used by the animals to perceive distance and direction. With echolocation, the bat or other animal can see not only where it is going but also how big another animal is, what kind of animal it is, and other features.

Bats

Microbats use echolocation to navigate and forage, often in total darkness. They generally emerge from their roosts in caves, attics, or trees at dusk and hunt for insects into the night. Their use of echolocation allows them to occupy a niche where there are often many insects (that come out at night since there are fewer predators then) and where there is less competition for food, and where there are fewer other species that may prey on the bats themselves.

Microbats generate ultrasound via the larynx and emit the sound through the open mouth or, much more rarely, the nose. The latter is most pronounced in the horseshoe bats (*Rhinolophus spp.*). Microbat calls range in frequency from 14,000 to well over 100,000 Hz, mostly beyond the range of the human ear (typical human hearing range is considered to be from 20 Hz to 20,000 Hz). Bats may estimate the elevation of targets by interpreting the interference patterns caused by the echoes reflecting from the tragus, a flap of skin in the external ear. There are two hypotheses about the evolution of echolocation in bats. The first suggests that laryngeal echolocation evolved twice in Chiroptera, once in Yangochiroptera and once in the rhinolophoids. The second proposes that laryngeal echolocation had a single origin in Chiroptera, was subsequently lost in the family Pteropodidae (all megabats), and later evolved as a system of tongue-clicking in the genus *Rousettus*.

Individual bat species echolocate within specific frequency ranges that suit their environment and prey types. This has sometimes been used by researchers to identify bats flying in an area simply by recording their calls with ultrasonic recorders known as 'bat detectors'. However echolocation calls are not always species specific and some bats overlap in the type of calls they use so recordings of echolocation calls cannot be used to identify all bats. In recent years researchers in several countries have developed 'bat call libraries' that contain recordings of local bat species that have been identified known as 'reference calls' to assist with identification.

Since the 1970s there has been an ongoing controversy among researchers as to whether bats use a form of processing known from radar termed coherent cross-correlation. Coherence means that the phase of the echolocation signals is used by the bats, while cross-correlation just implies that the outgoing signal is compared with the returning echoes in a running process. Today most - but not all - researchers believe that they use cross-correlation, but in an incoherent form, termed a filter bank receiver.

When searching for prey they produce sounds at a low rate (10-20/sec). During the search phase the sound emission is coupled to respiration, which is again coupled to the wingbeat. This coupling appears to dramatically conserve energy as there is little to no additional energetic cost of echolocation to flying bats (Speakman and Racey 1991). After detecting a potential prey item, microbats increase the rate of pulses, ending with the terminal buzz, at rates as high as 200/sec. During approach to a detected target, the duration of the sounds is gradually decreased, as is the energy of the sound.

Calls and Ecology

Bats belonging to the suborder Microchiroptera (microbats) occupy a diverse set of ecological conditions - they can be found living in environments as different as Europe and Madagascar, and hunting for food sources as different as insects, frogs, nectar, fruit, and blood. Additionally, the characteristics of an echolocation call are adapted to the particular environment, hunting behaviour, and food source of the particular bat. However, this adaptation of echolocation calls to ecological factors is constrained by the phylogenetic relationship of the bats, leading to a process known as descent with modification, and resulting in the diversity seen in the Microchiropteran suborder today. (Jones and Teeling 2006; Grinnell 1995; Zupanc 2004; Fenton 1995; Neuweiler 2003; Simmons and Stein 1980)

Acoustic Features

Describing the diversity of bat echolocation calls requires examination of the frequency and temporal features of the calls. It is the variations in these aspects that produce echolocation calls suited for different acoustic environments and hunting behaviours. (Fenton 2005; Jones and Teeling 2006; Zupanc 2004; Simmons and Stein 1980; Hiryu et al. 2007)

- Frequency Modulation and Constant Frequency: Echolocation calls can be composed of two different types of frequency structures: frequency modulated (FM) sweeps, and constant frequency (CF) tones. A particular call can consist of one, the other, or both structures. An FM sweep is a broadband signal – that is, it contains a downward sweep through a range of frequencies. A CF tone is a narrowband signal: the sound stays constant at one frequency throughout its duration.
- Intensity: Echolocation calls have been measured at intensities anywhere between 60 and 140 decibels(Surlykee et al. 2008). Certain microbat species can modify their call intensity mid-call, lowering the intensity as they approach objects that reflect sound strongly. This prevents the returning echo from deafening the bat (Hiryu et al. 2007). Additionally, the so-called "whispering bats" have adapted low-amplitude echolocation so that their prey, moths, which are able to hear echolocation calls, are less able to detect and avoid an oncoming bat (Fullard 1997).

- Harmonic composition: Calls can be composed of one frequency, or multiple frequencies comprising a harmonic series. In the latter case, the call is usually dominated by a certain harmonic ("dominant" frequencies are those present at higher intensities than other harmonics present in the call).
- Call duration: A single echolocation call (a call being a single continuous trace on a sound spectrogram, and a series of calls comprising a sequence or pass) can last anywhere from 0.2 to 100 milliseconds in duration, depending on the stage of prey-catching behaviour that the bat is engaged in. For example, the duration of a call usually decreases when the bat is in the final stages of prey capture – this enables the bat to call more rapidly without overlap of call and echo. Reducing duration comes at the cost of having less total sound available for reflecting off objects and being heard by the bat.
- Pulse interval: The time interval between subsequent echolocation calls (or pulses) determines two aspects of a bat's perception. First, it establishes how quickly the bat's auditory scene information is updated. For example, bats increase the repetition rate of their calls (that is, decrease the pulse interval) as they home in on a target. This allows the bat to get new information regarding the target's location at a faster rate when it needs it most. Secondly, the pulse interval determines the maximum range that bats can detect objects. This is because bats can only keep track of the echoes from one call at a time; as soon as they make another call they stop listening for echoes from the previously made call (Wilson and Moss 2004). For example, a pulse interval of 100 ms (typical of a bat searching for insects) allows sound to travel in air roughly 34 meters so a bat can only detect objects as far away as 17 meters (the sound has to travel out and back). With a pulse interval of 5 ms (typical of a bat in the final moments of a capture attempt), the bat can only detect objects up to 85 cm away. Therefore the bat constantly has to make a choice between getting new information updated quickly and detecting objects far away.

FM Signal Advantages

The major advantage conferred by an FM signal is extremely precise range discrimination, or localisation, of the target. J.A. Simmons demonstrated this effect with a series of elegant experiments that showed how bats using FM signals could distinguish between two separate targets even when the targets were less than half a millimeter apart. This amazing ability is due to the broadband sweep of the signal, which allows for better resolution of the time delay between the call and the returning echo, thereby improving the cross correlation of the two. Additionally, if harmonic frequencies are added to the FM signal, then this localisation becomes even more precise. (Jones and Teeling 2006; Zupanc 2004; Simmons and Stein 1980; Grinnell 1995)

One possible disadvantage of the FM signal is a decreased operational range of the call. Because the energy of the call is spread out among many frequencies, the distance at which the FM-bat can detect targets is limited (Fenton 1995). This is in part because any echo returning at a particular frequency can only be evaluated for a brief fraction of a millisecond, as the fast downward sweep of the call does not remain at any one frequency for long (Grinnell 1995).

CF Signal Advantages

The structure of a CF signal is adaptive in that it allows the CF-bat to detect both the velocity of a target, and the fluttering of a target's wings as Doppler shifted frequencies. A Doppler shift is an alteration in sound wave frequency, and is produced in two relevant situations: when the bat and its target are moving relative to each other, and when the target's wings are oscillating back and forth. CF-bats must compensate for Doppler shifts, lowering the frequency of their call in response to echoes of elevated frequency - this ensures that the returning echo remains at the frequency to which the ears of the bat are most finely tuned. The oscillation of a target's wings also produces amplitude shifts, which gives a CF-bat additional help in distinguishing a flying target from a stationary one. (Schnitzler and Flieger 1983; Zupanc 2004; Simmons and Stein 1980; Grinnell 1995; Neuweiler 2003; Jones and Teeling 2006)

Additionally, because the signal energy of a CF call is concentrated into a narrow frequency band, the operational range of the call is much greater than that of an FM signal. This relies on the fact that echoes returning within the narrow frequency band can be summed over the entire length of the call, which maintains a constant frequency for up to 100 milliseconds (Grinnell 1995; Fenton 1995)

Acoustic Environments of FM and CF Signals

- FM: An FM component is excellent for hunting prey while flying in close, cluttered environments. Two aspects of the FM signal account for this fact: the precise target localisation conferred by the broadband signal, and the short duration of the call. The first of these is essential because in a cluttered environment, the bats must be able to resolve their prey from large amounts of background noise. The 3D localisation abilities of the broadband signal enable the bat to do exactly that, providing it with what Simmons and Stein (1980) call a "clutter rejection strategy." This strategy is further improved by the use of harmonics, which, as previously stated, enhance the localisation properties of the call. The short duration of the FM call is also best in close, cluttered environments because it enables the bat to emit many calls extremely rapidly without overlap. This means that the bat can get an almost continuous stream of information – essential when objects are close, because they will pass by quickly – without confusing which echo corresponds to which call. (Neuweiler 2003; Simmons and Stein 1980; Jones and Teeling 2006; Fenton 1995)

- CF: A CF component is often used by bats hunting for prey while flying in open, clutter-free environments, or by bats that wait on perches for their prey to appear. The success of the former strategy is due to two aspects of the CF call, both of which confer excellent prey-detection abilities. First, the greater working range of the call allows bats to detect targets present at great distances – a common situation in open environments. Second, the length of the call is also suited for targets at great distances: in this case, there is a decreased chance that the long call will overlap with the returning echo. The latter strategy is made possible by the fact that the long, narrowband call allows the bat to detect Doppler shifts, which would be produced by an insect moving either towards or away from a perched bat. (Neuweiler 2003; Simmons and Stein 1980; Jones and Teeling 2006; Fenton 1995)

Neural Mechanisms in the Brain

Because bats use echolocation to orient themselves and to locate objects, their auditory systems are adapted for this purpose, highly specialised for sensing and interpreting the stereotyped echolocation calls characteristic of their own species. This specialisation is evident from the inner ear up to the highest levels of information processing in the auditory cortex.

Inner Ear and Primary Sensory Neurons

Both CF and FM bats have specialised inner ears which allow them to hear sounds in the ultrasonic range, far outside the range of human hearing. Although in most other aspects, the bat's auditory organs are similar to those of most other mammals, certain bats (horseshoe bats, *Rhinolophus spp.* and the moustached bat, *Pteronotus parnelii*) with a constant frequency (CF) component to their call (known as high duty cycle bats) do have a few additional adaptations for detecting the predominant frequency (and harmonics) of the CF vocalisation. These include a narrow frequency "tuning" of the inner ear organs, with an especially large area responding to the frequency of the bat's returning echoes (Neuweiler 2003).

The basilar membrane within the cochlea contains the first of these specialisations for echo information processing. In bats that use CF signals, the section of membrane that responds to the frequency of returning echoes is much larger than the region of response for any other frequency. For example, in *Rhinolophus ferrumequinum*, the horseshoe bat, there is a disproportionately lengthened and thickened section of the membrane that responds to sounds around 83 kHz, the constant frequency of the echo produced by the bat's call. This area of high sensitivity to a specific, narrow range of frequency is known as an "acoustic fovea" (Schuller and Pollack 1979).

Odontocetes (toothed whales and dolphins) have similar cochlear specialisations to those found in bats. Odontocetes also have the highest neural investment of any cochleae reported to date with ratios of greater than 1500 ganglion cells/mm of basilar membrane.

Further along the auditory pathway, the movement of the basilar membrane results in the stimulation of primary auditory neurons. Many of these neurons are specifically "tuned" (respond most strongly) to the narrow frequency range of returning echoes of CF calls. Because of the large size of the acoustic fovea, the number of neurons responding to this region, and thus to the echo frequency, is especially high (Carew 2001).

Inferior Colliculus

In the Inferior colliculus, a structure in the bat's midbrain, information from lower in the auditory processing pathway is integrated and sent on to the auditory cortex. As George Pollak and others showed in a series of papers in 1977, the interneurons in this region have a very high level of sensitivity to time differences, since the time delay between a call and the returning echo tells the bat its distance from the target object. Especially interesting is that while most neurons respond more quickly to stronger stimuli, collicular neurons maintain their timing accuracy even as signal intensity changes.

These interneurons are specialised for time sensitivity in several ways. First, when activated, they generally respond with only one or two action potentials. This short duration of response allows their action potentials to give a very specific indication of the exact moment of the time when the stimulus arrived, and to respond accurately to stimuli that occur close in time to one another. In addition, the neurons have a very low threshold of activation – they respond quickly even to weak stimuli. Finally, for FM signals, each interneuron is tuned to a specific frequency within the sweep, as well as to that same frequency in the following echo. There is specialisation for the CF component of the call at this level as well. The high proportion of neurons responding to the frequency of the acoustic fovea actually increases at this level (Carew 2001, Pollak 1977, Zupanc 2004).

Auditory Cortex

The auditory cortex in bats is quite large in comparison with other mammals (Anderson 1995). Various characteristics of sound are processed by different regions of the cortex, each providing different information about the location or movement of a target object. Most of the existing studies on information processing in the auditory cortex of the bat have been done by Nobuo Suga on the mustached bat, *Pteronotus parnellii*. This bat's call has both CF tone and FM sweep components.

Suga and his colleagues have shown that the cortex contains a series of "maps" of auditory information, each of which is organised systematically based on characteristics of sound such as frequency and amplitude. The neurons in these areas respond only to a specific combination of frequency and timing (sound-echo delay), and are known as combination-sensitive neurons.

The systematically organised maps in the auditory cortex respond to various aspects of the echo signal, such as its delay and its velocity. These regions are composed of "combination sensitive" neurons that require at least two specific stimuli to elicit a response. The neurons vary systematically across the maps, which are organised by acoustic features of the sound and can be two dimensional. The different features of the call and its echo are used by the bat to determine important characteristics of their prey.

- *FM-FM area:* This region of the cortex contains FM-FM combination-sensitive neurons. These cells respond only to the combination of two FM sweeps: a call and its echo. The neurons in the FM-FM region are often referred to as "delay-tuned," since each responds to a specific time delay between the original call and the echo, in order to find the distance from the target object (the range). Each neuron also shows specificity for one harmonic in the original call and a different harmonic in the echo. The neurons within the FM-FM area of the cortex of *Pteronotus* are organised into columns, in which the delay time is constant vertically but increases across the horizontal plane. The result is that range is encoded by location on the cortex, and increases systematically across the FM-FM area (Suga et al. 1975, Suga et al. 1979, Neuweiler 2003, Carew 2001).

- CF-CF area: Another kind of combination-sensitive neuron is the CF-CF neuron. These respond best to the combination of a CF call containing two given frequencies – a call at 30 kHz (CF1) and one of its additional harmonics around 60 or 90 kHz (CF2 or CF3) – and the corresponding echoes. Thus, within the CF-CF region, the changes in echo frequency caused by the Doppler shift can be compared to the frequency of the original call to calculate the bat's velocity relative to its target object. As in the FM-FM area, information is encoded by its location within the map-like organisation of the region. The CF-CF area is first split into the distinct CF1-CF2 and CF1-CF3 areas. Within each area, the CF1 frequency is organised on an axis, perpendicular to the CF2 or CF3 frequency axis. In the resulting grid, each neuron codes for a certain combination of frequencies that is indicative of a specific velocity (Suga et al. 1975, Suga et al. 1987, Carew 2001).
- DSCF area: This large section of the cortex is a map of the acoustic fovea, organised by frequency and by amplitude. Neurons in this region respond to CF signals that have been Doppler shifted (in other words, echoes only) and are within the same narrow frequency range to which the acoustic fovea responds. For *Pteronotus*, this is around 61 kHz. This area is organised into columns, which are arranged radially based on frequency. Within a column, each neuron responds to a specific combination of frequency and amplitude. Suga's studies have indicated that this brain region is necessary for frequency discrimination (Suga et al. 1975, Suga et al. 1987, Carew 2001).

Toothed Whales

Biosonar is valuable to Toothed whales (suborder odontoceti), including dolphins, porpoises, river dolphins, killer whales and sperm whales, because they live in an underwater habitat that has favourable acoustic characteristics and where vision is extremely limited in range due to absorption or turbidity.

Cetacean evolution consisted of three main radiations. Throughout the middle and late Eocene periods (49-31.5 million years ago), archaeocetes, primitive toothed Cetacea that arose from terrestrial mammals with the creation of aquatic adaptations, were the only known archaic Cetacea. These primitive aquatic mammals did not posses the ability to echolocate, although they did have slightly adapted underwater hearing. The morphology of acoustically isolated ear bones in basilosaurid archaeocetes indicates that this order had directional hearing underwater at low to mid frequencies by the late middle Eocene. However, with the extinction of archaeocete at the onset of the Oligocene, two new lineages in the early Oligocene period (31.5-28 million years ago) compromised a second radiation. These early mysticete (baleen whales) and odontocete can be dated back to the middle Oligocene in New Zealand. Based on past phylogenies, it has been found that the evolution of odontocetes is monophyletic, suggesting that echolocation evolved only once 36 to 34 million years ago. Dispersal rates routes of early odontocetes included transoceanic travel to new adaptive zones. The third radiation occurred later in the Neogene, when present dolphins and their relatives evolved to be the most common species in the modern sea.

The evolution of echolocation could be attributed several theories. There are two proposed drives for the hypotheses of cetacean radiation, one biotic and the other abiotic in nature. The first, adaptive radiation, is the result of a rapid divergence into new adaptive zones. This results in diverse, ecologically different clades that are incomparable. Clade Neocete (crown cetacean) has been characterised by an evolution from archaeocetes and a dispersion across the world's oceans, and even estuarites and rivers. These ecological opportunities were the result of abundant dietary resources with low competition for hunting. This hypothesis of lineage diversification, however, can be unconvincing due to a lack of support for rapid speciation early in cetacean history. A second, more abiotic drive is more supported. Physical restructuring of the oceans has played a role in echolocation radiation. This was a result of global climate change at the Eocene-Oligocene boundary; from a greenhouse to an icehouse world. Tectonic openings created the emergence of the Southern ocean with a free flowing Antarctic Circumpolar current (Fordyce 1980, 2003; Lindberg & Pyenson 2007; Steeman et al. 2009). These events allowed for a selection regime characterised by the ability to locate and capture prey in turbid river waters, or allow odontocetes to invade and feed at depths below the photic zone. Further studies have found that echolocation below the photic zone could have been a predation adaptation to diel migrating cephalopods. Since its advent, there has been adaptive radiation especially in the Delphinidae family (dolphins) in which echolocation has become extremely derived.

One specific type of echolocation, narrow-band high frequency (NBHF) clicks, evolved at least four times in groups of odontocetes, including pygmy sperm whale (Kogiidae) and porpoise (Phocoenidae) families, *Pontoporia blainvillei*, the genus *Cephalorhynchus*, and part of the genus *Lagenorhynchus*. These high frequency clicks likely evolved as adaptation of predator avoidance, as they inhabit areas that have many killer whales and the signals are inaudible to killer whales due to the absence of energy below 100 kHz. Another reason for variation in echolocation frequencies is habitat. Shallow waters, where many of these species live, tend to have more debris; a more directional transmission reduces clutter in reception.

Toothed whales emit a focused beam of high-frequency clicks in the direction that their head is pointing. Sounds are generated by passing air from the bony nares through the phonic lips. These sounds are reflected by the dense concave bone of the cranium and an air sac at its base. The focused beam is modulated by a large fatty organ known as the 'melon'. This acts like an acoustic lens because it is composed of lipids of differing densities. Most toothed whales use clicks in a series, or click train, for echolocation, while the sperm whale may produce clicks individually. Toothed whale whistles do not appear to be used in echolocation. Different rates of click production in a click train give rise to the familiar barks, squeals and growls of the bottlenose dolphin. A click train with a repetition rate over 600 per second is called a burst pulse. In bottlenose dolphins, the auditory brain response resolves individual clicks up to 600 per second, but yields a graded response for higher repetition rates.

It has been suggested that some smaller toothed whales may have their tooth arrangement suited to aid in echolocation. The placement of teeth in the jaw of a bottlenose dolphin, as an example, are not symmetrical when seen from a vertical plane, and this asymmetry could possibly be an aid in the dolphin sensing if echoes from its biosonar are coming from one side or the other. However, this idea lacks experimental support.

Echoes are received using complex fatty structures around the lower jaw as the primary reception path, from where they are transmitted to the middle ear via a continuous fat body (Ketten 1992,2000). Lateral sound may be received though fatty lobes surrounding the ears with a similar density to water. Some researchers believe that when they approach the object of interest, they protect themselves against the louder echo by quietening the emitted sound. In bats this is known to happen, but here the hearing sensitivity is also reduced close to a target.

Before the echolocation abilities of "porpoises" were officially discovered, Jacques Yves Cousteau suggested that they might exist. In his first book, *The Silent World* (1953, pp. 206–207), he reported that his research vessel, the *Élie Monier,* was heading to the Straits of Gibraltar and noticed a group of porpoises following them. Cousteau changed course a few degrees off the optimal course to the center of the strait, and the porpoises followed for a few minutes, then diverged toward mid-channel again. It was obvious that they knew where the optimal course lay, even if the humans didn't. Cousteau concluded that the cetaceans had something like sonar, which was a relatively new feature on submarines.

Oilbirds and Swiftlets

Oilbirds and some species of swiftlet are known to use a relatively crude form of echolocation compared to that of bats and dolphins. These nocturnal birds emit calls while flying and use the calls to navigate through trees and caves where they live.

Shrews and Tenrecs

Terrestrial mammals other than bats known to echolocate include two genera (*Sorex* and *Blarina*) of shrews and the tenrecs of Madagascar. These include the wandering shrew (*Sorex vagrans*), the common or Eurasian shrew (*Sorex araneus*), and the short-tailed shrew (*Blarina brevicauda*). The nature of shrew sounds unlike those of bats are low amplitude, broadband, multi-harmonic and frequency modulated. They contain no 'echolocation clicks' with reverberations and would seem to be used for simple, close range spatial orientation. In contrast to bats, shrews use echolocation only to investigate their habitat rather than additionally to pinpoint food.

Snake

Snakes are elongate, legless, carnivorous reptiles of the suborder Serpentes that can be distinguished from legless lizards by their lack of eyelids and external ears. Like all squamates, snakes are ectothermic, amniote vertebrates covered in overlapping scales. Many species of snakes have skulls with many more joints than their lizard ancestors, enabling them to swallow prey much larger than their heads with their highly mobile jaws. To accommodate their narrow bodies, snakes' paired organs (such as kidneys) appear one in front of the other instead of side by side, and most have only one functional lung. Some species retain a pelvic girdle with a pair of vestigial claws on either side of the cloaca.

Figure 1: *Coast garter snake*

Living snakes are found on every continent except Antarctica and on most islands. Fifteen families are currently recognised, comprising 456 genera and over 2,900 species. They range in size from the tiny, 10 cm-long thread snake to pythons and anacondas of up to 7.6 meters (25 ft) in length. The fossil species *Titanoboa cerrejonensis* was 15 meters (49 ft) long. Snakes are thought to have evolved from either burrowing or aquatic lizards during the mid-Cretaceous period, and the earliest known fossils date to around 112 Ma ago. The diversity of modern snakes appeared during the Paleocene period (*c* 66 to 56 Ma ago).

Most species are nonvenomous and those that have venom use it primarily to kill and subdue prey rather than for self-defense. Some possess venom potent enough to cause painful injury or death to humans. Nonvenomous snakes either swallow prey alive or kill by constriction.

Etymology

The English word *snake* comes from Old English *snaca*, itself from Proto-Germanic **snak-an-* (cf. German *Schnake* "ring snake," Swedish *snok* "grass snake"), from Proto-Indo-European root **(s)nçg-o-* "to crawl, creep," which also gave *sneak* as well as Sanskrit *nâgá* "snake." The word ousted *adder*, as *adder* went on to narrow in meaning, though in Old English *nœddre* was the general word for snake. The other term, *serpent*, is from French, ultimately from Indo-European **serp-* (to creep), which also gave Greek *érpo* (åñðù) "I crawl."

Evolution

The fossil record of snakes is relatively poor because snake skeletons are typically small and fragile, making fossilisation uncommon. Fossils readily identifiable as snakes (though often retaining hind limbs) first appear in the fossil record during the

Cretaceous period. The earliest known snake fossils come from sites in Utah and Algeria, represented by the genera *Coniophis* and *Lapparentophis*, respectively. These fossil sites have been tentatively dated to the Albian or Cenomanian age of the late Cretaceous, between 112 and 94 Ma ago. However, an even older age has been suggested for one of the Algerian sites, which may be as old as the Aptian, 125-112 Ma ago.

Based on comparative anatomy, there is consensus that snakes descended from lizards. Pythons and boas—primitive groups among modern snakes—have vestigial hind limbs: tiny, clawed digits known as anal spurs, which are used to grasp during mating. The Leptotyphlopidae and Typhlopidae groups also possess remnants of the pelvic girdle, sometimes appearing as horny projections when visible.

Frontal limbs are nonexistent in all known snakes. This is caused by the evolution of Hox genes, controlling limb morphogenesis. The axial skeleton of the snakes' common ancestor, like most other tetrapods, had regional specialisations consisting of cervical (neck), thoracic (chest), lumbar (lower back), sacral (pelvic), and caudal (tail) vertebrae. Early in snake evolution, the Hox gene expression in the axial skeleton responsible for the development of the thorax became dominant. As a result, the vertebrae anterior to the hindlimb buds (when present) all have the same thoracic-like identity (except from the atlas, axis, and 1–3 neck vertebrae). In other words, most of a snake's skeleton is an extremely extended thorax. Ribs are found exclusively on the thoracic vertebrae. Neck, lumbar and pelvic vertebrae are very reduced in number (only 2–10 lumbar and pelvic vertebrae are present), while only a short tail remains of the caudal vertebrae. However, the tail is still long enough to be of important use in many species, and is modified in some aquatic and tree-dwelling species.

Modern snakes greatly diversified during the Paleocene. This occurred alongside the adaptive radiation of mammals, following the extinction of (non-avian) dinosaurs. The colubrids, one of the more common snake groups, became particularly diverse due to preying on rodents, an especially successful mammal group. There are over 2,900 species of snakes ranging as far northward as the Arctic Circle in Scandinavia and southward through Australia and Tasmania. Snakes can be found on every continent (with the exception of Antarctica), in the sea, and as high as 16,000 feet (4,900 m) in the Himalayan Mountains of Asia. There are numerous islands from which snakes are absent, such as Ireland, Iceland, and New Zealand.

Origins

The origin of snakes remains an unresolved issue. There are two main hypotheses competing for acceptance.

Burrowing Lizard Hypothesis

There is fossil evidence to suggest that snakes may have evolved from burrowing lizards, such as the varanids (or a similar group) during the Cretaceous Period. An early fossil snake, *Najash rionegrina*, was a two-legged burrowing animal with a sacrum, and was fully terrestrial. One extant analog of these putative ancestors is the earless monitor *Lanthanotus* of Borneo (though it also is semiaquatic). Subterranean species evolved bodies streamlined for burrowing, and eventually lost their limbs. According to this hypothesis, features such as the transparent, fused eyelids (brille) and loss of external ears evolved to cope with fossorial difficulties, such as scratched corneas and dirt in the ears. Some primitive snakes are known to have possessed hindlimbs, but their pelvic bones lacked a direct connection to the vertebrae. These include fossil species like *Haasiophis*, *Pachyrhachis* and *Eupodophis*, which are slightly older than *Najash*.

An alternative hypothesis, based on morphology, suggests the ancestors of snakes were related to mosasaurs—extinct aquatic reptiles from the Cretaceous—which in turn are thought to have derived from varanid lizards. According to this hypothesis, the fused, transparent eyelids of snakes are thought to have evolved to combat marine conditions (corneal water loss through osmosis), and the external ears were lost through disuse in an aquatic environment. This ultimately lead to an animal similar to today's sea snakes. In the Late Cretaceous, snakes recolonised land, and continued to diversify into today's snakes. Fossilised snake remains are known from early Late Cretaceous marine sediments, which is consistent with this hypothesis; particularly so, as they are older than the terrestrial *Najash rionegrina*. Similar skull structure, reduced or absent limbs, and other anatomical features found in both mosasaurs and snakes lead to a positive cladistical correlation, although some of these features are shared with varanids.

Genetic studies in recent years have indicated snakes are not as closely related to monitor lizards as was once believed—and therefore not to mosasaurs, the proposed ancestor in the aquatic scenario of their evolution. However, more evidence links mosasaurs to snakes than to varanids. Fragmented remains found from the Jurassic and Early Cretaceous indicate deeper fossil records for these groups, which may potentially refute either hypothesis.

Taxonomy

All modern snakes are grouped within the suborder *Serpentes* in Linnean taxonomy, part of the order Squamata, though their precise placement within squamates is controversial.

There are two infraorders of *Serpentes*: Alethinophidia and Scolecophidia. This separation is based on morphological characteristics and mitochondrial DNA sequence similarity. Alethinophidia is sometimes split into Henophidia and Caenophidia, with the latter consisting of "colubroid" snakes (colubrids, vipers, elapids, hydrophiids, and attractaspids) and acrochordids, while the other alethinophidian families comprise Henophidia. While not extant today, the Madtsoiidae, a family of giant, primitive, python-like snakes, was around until 50,000 years ago in Australia, represented by genera such as *Wonambi.*

There are numerous debates in the systematics within the group. For instance, many sources classify Boidae and Pythonidae as one family, while some keep the Elapidae and Hydrophiidae (sea snakes) separate for practical reasons despite their extremely close relation. Recent molecular studies support the monophyly of the clades of modern snakes, scolecophidians, typhlopids + anomalepidids, alethinophidians, core alethinophidians, uropeltids (*Cylindrophis, Anomochilus,* uropeltines), macrostomatans, booids, boids, pythonids and caenophidians.

Biology

Skeleton: The skeleton of most snakes consists solely of the skull, hyoid, vertebral column, and ribs, though henophidian snakes retain vestiges of the pelvis and rear limbs. The skull of the snake consists of a solid and complete braincase, to which many of the other bones are only loosely attached, particularly the highly mobile jaw bones, which facilitate manipulation and ingestion of large prey items. The left and right sides of the lower jaw are joined only by a flexible ligament at the anterior tips, allowing them to separate widely, while the posterior end of the lower jaw bones articulate with a quadrate bone, allowing further mobility. The bones of the mandible and quadrate bones can also pick up ground borne vibrations. The hyoid is a small bone located posterior and ventral to the skull, in the 'neck' region, which serves as an attachment for muscles of the snake's tongue, as it does in all other tetrapods.

The vertebral column consists of anywhere between 200 to 400 (or more) vertebrae. Tail vertebrae are comparatively few in number (often less than 20% of the total) and lack ribs, while body vertebrae each have two ribs articulating with them. The vertebrae have projections that allow for strong muscle attachment enabling locomotion without limbs. Autotomy of the tail, a feature found in some lizards is absent in most snakes. Caudal autotomy in snakes is rare and is intervertebral, unlike that in lizards, which is intravertebral—that is, the break happens along a predefined fracture plane present on a vertebra.

In some snakes, most notably boas and pythons, there are vestiges of the hindlimbs in the form of a pair of pelvic spurs. These small, claw-like protrusions on each side of the cloaca are the external portion of the vestigial hindlimb skeleton, which includes the remains of an ilium and femur.

Internal Organs

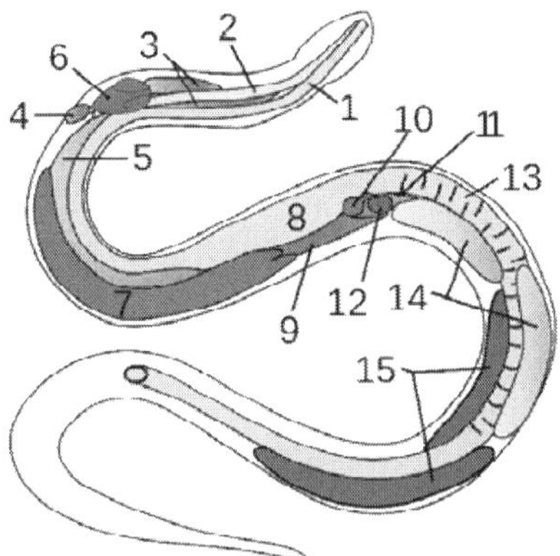

Figure: *Anatomy of a snake. 1 esophagus, 2 trachea, 3 tracheal lungs, 4 rudimentary left lung, 5 right lung, 6 heart, 7 liver, 8 stomach, 9 air sac, 10 gallbladder, 11 pancreas, 12 spleen, 13 intestine, 14 testicles, 15 kidneys.*

The snake's heart is encased in a sac, called the *pericardium*, located at the bifurcation of the bronchi. The heart is able to move around, however, owing to the lack of a diaphragm. This adjustment protects the heart from potential damage when large ingested prey is passed through the esophagus. The spleen is attached to the gall bladder and pancreas and filters the blood. The thymus gland is located in fatty tissue above the heart and is responsible for the generation of immune cells in the blood. The cardiovascular system of snakes is also unique for the presence of a renal portal system in which the blood from the snake's tail passes through the kidneys before returning to the heart.

The vestigial left lung is often small or sometimes even absent, as snakes' tubular bodies require all of their organs to be long and thin. In the majority of species, only one lung is functional. This lung contains a vascularised anterior portion and a posterior portion that does not function in gas exchange. This 'saccular lung' is used for hydrostatic purposes to adjust buoyancy in some aquatic snakes and its function remains unknown in terrestrial species. Many organs that are paired, such as kidneys or reproductive organs, are staggered within the body, with one located ahead of the other.

Snakes have no lymph nodes.

Size

The now extinct *Titanoboa cerrejonensis* snakes found were 12–15 meters (39–49 ft) in length. By comparison, the largest extant snakes are the reticulated python, which measures about 9 meters (30 ft) long, and the anaconda, which measures about 7.5 meters (25 ft) long and is considered the heaviest snake on Earth.

At the other end of the scale, the smallest extant snake is *Leptotyphlops carlae*, with a length of about 10 centimeters (4 in). Most snakes are fairly small animals, approximately 1 meter (3 feet) in length.

Snake Scales

Snakes, like other reptiles, have a skin covered in scales. Snakes are entirely covered with scales or scutes of various shapes and sizes. Scales protect the body of the snake, aid it in locomotion, allow moisture to be retained within, alter the surface characteristics such as roughness to aid in camouflage, and in some cases even aid in prey capture (such as Acrochordus). The simple or complex colouration patterns (which help in camouflage and anti-predator display) are a property of the underlying skin, but the folded nature of scaled skin allows bright skin to be concealed between scales then revealed in order to startle predators.

Figure: *Elaborately shaped scales on the head of a Vine snake, Ahaetulla nasuta.*

Scales have been modified over time to serve other functions such as 'eyelash' fringes, and protective covers for the eyes with the most distinctive modification being the *rattle* of the North American rattlesnakes.

Snakes periodically moult their scaly skins and acquire new ones. This permits replacement of old worn out skin, disposal of parasites and is thought to allow the snake to grow. The arrangement of scales is used to identify snake species.

Snakes have been part and parcel of culture and religion. Vivid scale patterns have been thought to have influenced early art. The use of snake-skin in manufacture of purses, apparel and other articles led to large-scale killing of snakes, giving rise to advocacy for use of artificial snake-skin. Snake scales are also to be found as motifs in fiction, video games and films.

Functions of Scales

The scales of a snake primarily serve to reduce friction as it moves, since friction is the major source of energy loss in snake locomotion.

The ventral (or belly) scales, which are large and oblong, are especially low-friction, and some arboreal species can use the edges to grip branches. Snake skin and scales help retain moisture in the animal's body. Snakes pick up vibrations from both the air and the ground, and can differentiate the two, using a complex system of internal resonances (perhaps involving the scales).

Figure: *Rainbow boas get their name from the colouration of their scales caused by iridescence.*

Morphology of scales

Snake scales are formed by the differentiation of the snake's underlying skin or epidermis. Each scale has an outer surface and an inner surface. The skin from the inner surface hinges back and forms a free area which overlaps the base of the next scale which emerges below this scale. A snake hatches with a fixed number of scales. The scales do not increase in number as the snake matures nor do they reduce in number over time. The scales however grow larger in size and may change shape with each moult. Snakes have smaller scales around the mouth and sides of the body which allow expansion so that a snake can consume prey of much larger width than itself. Snake scales are made of keratin, the same material that hair and fingernails are made of. They are cool and dry to touch.

Surface and Shape

Snake scales are of different shapes and sizes. Snake scales may be granular, have a smooth surface or have a longitudinal ridge or keel on it. Often, snake scales have pits, tubercles and other fine structures which may be visible to the naked eye or under a microscope. Snake scales may be modified to form fringes, as in the case of the Eyelash Bush Viper, *Atheris ceratophora*, or rattles as in the case of the rattlesnakes of North America. Certain primitive snakes such as boas, pythons and certain advanced snakes such as vipers have small scales arranged irregularly on the head. Other more advanced snakes have special large symmetrical scales on the head called *shields* or *plates*. Snake scales occur in variety of shapes. They may be cycloid as in family.

Typhlopidae, long and pointed with pointed tips, as in the case of the Green Vine Snake *Ahaetulla nasuta*, broad and leaf-like, as in the case of green pit vipers *Trimeresurus* spp or as broad as they are long, for example, as in Rat snake *Ptyas mucosus*.

In some cases, scales may be keeled weakly or strongly as in the case of the Buff-striped keelback *Amphiesma stolatum*. They may have bidentate tips as in some spp of *Natrix*. Some snakes, such as the Short Seasnake *Lapemis curtus*, may have spinelike and juxtaposed scales while others may have large and non-overlapping knobs as in the case of the Javan Mudsnake *Xenodermis javanicus*.

Another example of differentiation of snake scales is a transparent scale called the *brille* or *spectacle* which covers the eye of the snake. The brille is often referred to as a fused eyelid. It is shed as part of the old skin during moulting.

Rattles

The most distinctive modification of the snake scale is the *rattle* of rattlesnakes, such as those of the genera *Crotalus* and *Sistrurus*. The rattle is made up of a series of loosely linked, interlocking chambers that when shaken, vibrate against one another to create the warning signal of a rattlesnake. Only the bottom button is firmly attached to the tip of the tail.

At birth, a rattlesnake hatchling has only a small button or 'primordial rattle' which is firmly attached to the tip of the tail. The first segment is added when the hatchling sheds its skin for the first time. A new section is added each time the skin is shed until a rattle is formed. The rattle grows as the snake ages but segments are also prone to breaking off and hence the length of a rattle is not a reliable indicator of the age of a snake.

Colour

Scales, more specifically, mostly consist of hard beta keratins which are basically transparent. The colours of the scale are due to pigments in the inner layers of the skin and not due to the scale material itself. Scales are hued for all colours in this manner except for blue and green. Blue is caused by the ultrastructure of the scales. By itself, such a scale surface diffracts light and gives a blue hue, while, in combination with yellow from the inner skin it gives a beautiful iridescent green.

Some snakes have the ability to change the hue of their scales slowly. This is typically seen in cases where the snake becomes lighter or darker with change in season. In some cases, this change may take place between day and night.

Ecdysis

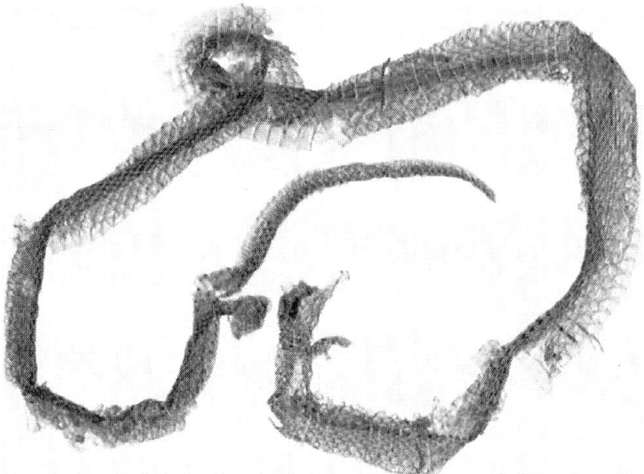

Figure: *Empty skin (exuvia) of a Grass Snake (Natrix natrix) whose length exceeds one metre.*

The shedding of scales is called *ecdysis*, or, in normal usage *moulting* or *sloughing*. In the case of snakes, the complete outer layer of skin is shed in one layer. Snake scales are not discrete but extensions of the epidermis hence they are not shed separately, but are ejected as a complete contiguous outer layer of skin during each moult, akin to a sock being turned inside out.

Moulting serves a number of functions - firstly, the old and worn skin is replaced, secondly, it helps get rid of parasites such as mites and ticks. Renewal of the skin by moulting is supposed to allow growth in some animals such as insects, however this view has been disputed in the case of snakes.

Moulting is repeated periodically throughout a snake's life. Before a moult, the snake stops eating and often hides or moves to a safe place. Just before shedding, the skin becomes dull and dry looking and the eyes become cloudy or blue-coloured. The inner surface of the old outer skin liquefies. This causes the old outer skin to separate from the new inner skin. After a few days, the eyes clear and the snake "crawls" out of its old skin. The old skin breaks near the mouth and the snake wriggles out aided by rubbing against rough surfaces. In many cases the cast skin peels backward over the body from head to tail, in one piece like an old sock. A new, larger, and brighter layer of skin has formed underneath. An older snake may shed its skin only once or twice a year, but a younger, still-growing snake, may shed up to four times a year. The discarded skin gives a perfect imprint of the scale pattern and it is usually possible to identify the snake if this discard is reasonably complete and intact.

Arrangement of Scales

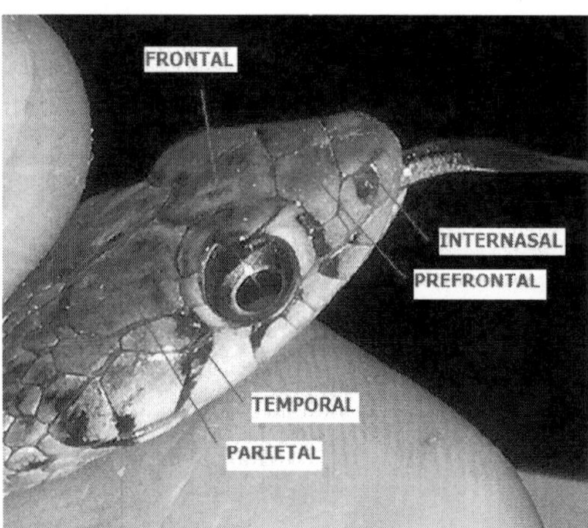

Figure: *Nomenclature of head scales (top view of head)*

Scale arrangements are important, not only for taxonomic utility, but also for forensic reasons and conservation of snake species. Excepting for the head, snakes have imbricate scales, overlapping like the tiles on a roof. Snakes have rows of scales along the whole or part of their length and also many other specialised scales, either singly or in pairs, occurring on the head and other regions of the body.

The dorsal (or body) scales on the snake's body are arranged in rows along the length of their bodies. Adjacent rows are diagonally offset from each other. Most snakes have an odd number of rows across the body though certain species have an even number of rows e.g. *Zaocys* spp. In the case of some aquatic and marine snakes, the scales are granular and the rows cannot be counted.

The number of rows range from ten in Tiger Ratsnake *Spilotes pullatus*; thirteen in *Dryocalamus, Liopeltis, Calamaria* and Asian coral snakes of genus *Calliophis*; 65 to 75 in pythons; 74 to 93 in *Kolpophis* and 130 to 150 in *Acrochordus*. The majority of the largest family of snakes, the *Colubridae* have 15, 17 or 19 rows of scales. The maximum number of rows are in mid-body and they reduce in count towards the head and on the tail.

Nomenclature of Scales

The various scales on a snake's head and body are indicated in the following paragraphs with annotated photographs of Buff-striped Keelback *Amphiesma stolata*, a common grass-snake of South Asia and a member of Colubridae, the largest snake family.

Head Scales

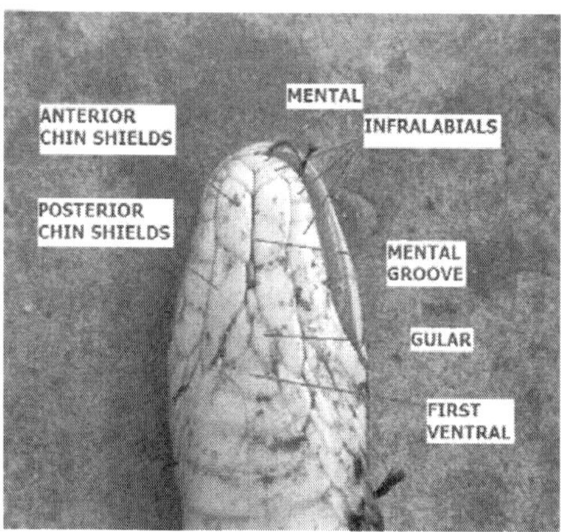

Figure: *Nomenclature of scales (underside view of head)*

Identification of cephalic scales is most conveniently begun with reference to the nostril which is easily identified on the snake. There are two scales enclosing the nostril which are called the nasals. In colubrids, the nostril lies between the nasals while in vipers it lies in the centre of a single nasal scale. The outer nasal (near the snout) is called the prenasal while the inner nasal (near the eye) is called the postnasal. Along the top of the snout connecting the nasals on both sides of the head are scales called internasals. Between the two prenasals is a scale at the tip of the snout called the rostral scale.

The scales around the eye are called circumorbital scales and are named as 'ocular' scales but with appropriate prefix. The ocular scale proper is a transparent scale covering the eye which is called the spectacle, brille or eyecap. The circumorbital scales towards the snout or the front are called preocular scales, those towards the rear are called postocular scales and those towards the upper or dorsal side are called as supraocular scales. Circumorbital scales towards the ventral or lower side, if any, are called as subocular scales. Between the preocular and the postnasal scales are one or two scales called as loreal scales. Loreal scales are absent in elapids.

The scales along the lips of the snake are called as labials. Those on the upper lip are called supralabials while those on the lower labial are called infralabials. Between the eyeballs on top of the head, adjacent to the supraoculars are the frontal scales. The prefrontal scales are the scales connected to the frontals towards the tip of the snout which are in contact with the internasals. They may have a scale in between them. The back of the top of the head has scales connected to the frontal scales called as the parietal scales. At the sides of the back of the head between the parietals above and the supralabials below are scales called temporal scales.

On the underside of the head, a snake has an anterior scale called as the mental scale. Connected to the mental scales and all along the lower jaws are the infralabials. Along the lower jaw connected to infralabials are a pair of shields called the anterior chin shields. Next to the anterior chin shields, further back along the jaw are another pair of shields called the posterior chin shields. In some texts the chinshields are referred to as *submaxillary* scales. Scales in the central or *throat* region, which are in contact with the first ventral scales of a snake's body and are flanked by the chin shields, are called gular scales. The mental groove is a longitudinal groove on the underside of the head between large, paired chin shields and smaller gular scales.

Body Scales

The scales on the body of the snake are called the dorsal or costal scales. Sometimes there is a special row of large scales along the top of the back of the snake, i.e., the uppermost row, called the vertebral scales. The enlarged scales on the belly of the snake are called ventral scales or gastrosteges. The number of ventral scales can be a guide to the species. In "advanced" (Caenophidian) snakes, the broad belly scales and rows of dorsal scales correspond to the vertebrae, allowing scientists to count the vertebrae without dissection.

Tail Scales

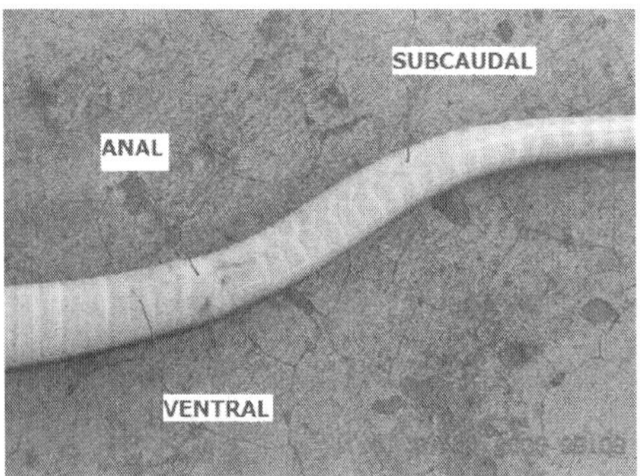

***Figure** 8: Nomenclature of scales (underside view of body),*

At the end of the ventral scales of the snake is an anal plate which protects the opening to the cloaca (a shared opening for waste and reproductive material to pass) on the underside near the tail. This anal scale may be single or divided into a pair.

The part of the body beyond the anal scale is considered to be the tail. Sometimes snakes have enlarged scales, either single or paired, under the tail; these are called subcaudals or urosteges. These subcaudals may be smooth or keeled as in *Bitis arietans somalica*. The end of the tail may simply taper into a tip (as in the case of most snakes), it may form a spine (as in *Acanthophis*), end in a bony spur (as in *Lachesis*), a rattle (as in *Crotalus*), or a rudder as seen in many sea snakes.

Glossary of Scales

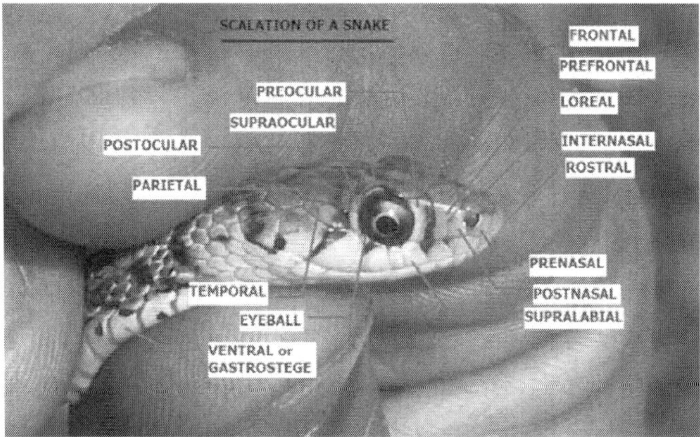

***Figure:** Nomenclature of scales (side view of head)*

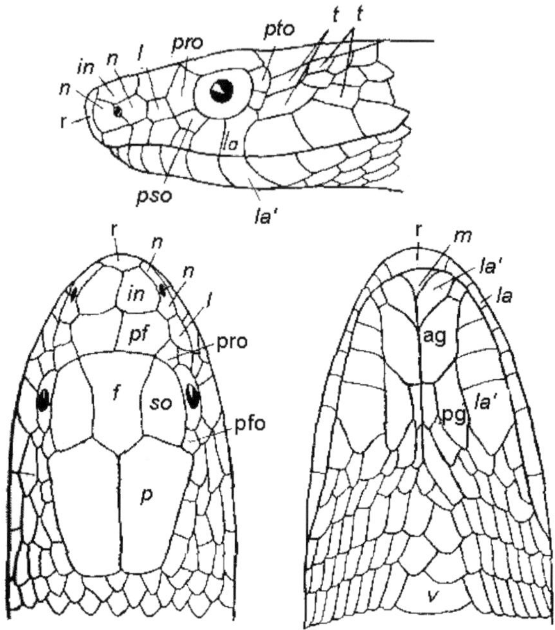

Figure 10: *Terminology of snake head shields explained with help of line diagrams of the head of Coluber ventromaculatus from Malcolm A. Smith (1943).*

Legend

ag - Anterior genials or Chin shields

f - Frontal

in - Internasal

l - Loreal

la - Supralabial

la' - Infralabial

m - Mental

n - Nasal

p - Parietal

pf - Prefrontal

pg - Posterior Genials or Chin shields

pro - Preocular

pso - Presubocular

pto - Postocular

r - Rostral

so - Supraocular

t - Anterior and Posterior Temporals

v - First Ventral

- Scales on the head.
  - Rostral.
  - Nasorostral.
  - Nasal.
    - Prenasal.
    - Postnasal.
    - Supranasal.
    - Fronto-nasal.
  - Internasal.
  - Brille, spectacle, ocular scale, eyecap.
  - Circumorbital.
    - Preocular.
    - Postocular.
    - Supraocular.
    - Subocular.
  - Loreal.
  - Interorbital, Intersupraocular.
  - Frontal.
  - Prefrontal.
  - Parietal.
  - Occipital.
  - Interoccipital.
  - Temporal.
  - Labial.
    - Supralabial, Upper labial.
    - Sublabial, Infralabial, Lower labial.
  - Mental or Symphysial.
  - Chin shield.
    - Anterior chin shield, Anterior genials.
    - Posterior chin shield, Posterior genials.
    - Intergeneial.
  - Gular.

- Scales on the body.
  - Dorsal.
  - Vertebral.
  - Ventral, Gastrostege.
- Scales on the tail.
  - Anal.
  - Subcaudal, Urostege.

Other Pertinent Terms

- Canthus, or *Canthus rostralis.*
- Mental groove.

Taxonomic Importance

Scales do not play an important role in distinguishing between the families but are important at generic and specific level. There is an elaborate scheme of nomenclature of scales. Scales patterns, by way of scale surface or texture, pattern and colouration and the division of the anal plate, in combination with other morphological characteristics, are the principal means of classifying snakes down to species level.

In certain areas in North America, where the diversity of snakes is not too large, easy keys based on simple identification of scales have been devised for the lay public to distinguish venomous snakes from non-venomous snakes. In other places with large biodiversity, such as Myanmar, publications caution that venomous and non-venomous snakes cannot be easily distinguished apart without careful examination.

The scales patterning may also be used for individual identification in field studies. Clipping of specific scales, such as the subcaudals, to mark individual snakes is a popular approach to population estimation by mark and recapture techniques.

Distinguishing between Venomous and Non-venomous Snakes

There is no simple way of differentiating a venomous snake from a non-venomous one merely by using a scale character. Finding out whether a snake is venomous or not is correctly done by identification of the species of a snake with the help of experts, or in their absence, close examination of the snake and using authoritative references on the snakes of the particular geographical region to identify it. Scale patterns help to indicate the species and from the references, it can be verified if the snake species is known to be venomous or not.

Figure: *Banded Krait (Bungarus fasciatus), an Elapid, with no loreal scale between nasal and pre-ocular scales.*

Species identification using scales requires a fair degree of knowledge about snakes, their taxonomy, snake-scale nomenclature as well as familiarity with and access to scientific literature. Distinguishing by using scale diagrams whether a snake is venomous or not in the field cannot be done in the case of uncaught specimens. It is not advisable to catch a snake to check whether it is venomous or not using scale diagrams. Most books or websites provide an array of traits of the local herpetofauna, other than scale diagrams, which help to distinguish whether a snake in the field is venomous or not.

In certain regions, presence or absence of certain scales may be a quick way to distinguish non-venomous and venomous snakes, but used with care and knowledge of exceptions. For example, in Myanmar, the presence or absence of loreal scales can be used to distinguish between relatively harmless Colubrids and lethally venomous Elapids. The rule of hand for this region is that the absence of a loreal scale between the nasal scale and pre-ocular scale indicates that the snake is an Elapid and hence lethal. This rule-of-thumb cannot be used without care as it cannot be applied to vipers, which have a large number of small scales on the head. A careful check would also be needed to exclude known poisonous members of the Colubrid family such as *Rhabdophis.*

In South Asia, it is advisable to take the snake which has bitten a person, if it has been killed, and carry it along to the hospital for possible identification by medical staff using scale diagrams so that an informed decision can be taken them as to whether and which anti-venom is to be administered. However, attempts to catch it or kill the venomous snake are not advised as the snake may bite more people.

Cultural Significance

Snakes have been a motif in human culture and religion and an object of dread and fascination all over the world. The vivid patterns of snake scales, such as the Gaboon Viper, both repel and fascinate the human mind. Such patterns have inspired dread and awe in humans from pre-historic times and these can be seen in the art prevalent to those times. Studies of fear imagery and psychological arousal indicate that snake scales are a vital component of snake imagery. Snake scales also appear to have affected Islamic art in the form of tessallated mosaic patterns which show great similarity to snake-scale patterns.

Snakeskin, with its highly periodic cross-hatch or grid patterns, appeals to people's aesthetics and have been used to manufacture many leather articles including fashionable accessories. The use of snakeskin has however endangered snake populations and resulted in international restrictions in trade of certain snake species and populations in the form of CITES provisions. Animal lovers in many countries now propagate the use of artificial snakeskin instead, which are easily produced from embossed leather, patterned fabric, plastics and other materials.

Snake scales occur as a motif regularly in computer action games. A snake scale was portrayed as a clue in the 1982 film *Blade Runner*. Snake scales also figure in popular fiction, such as the Harry Potter series (desiccated Boomslang skin is used as a raw material for concocting the Polyjuice potion), and also in teen fiction.

Moulting

Molting serves a number of functions. Firstly, the old and worn skin is replaced; secondly, it helps get rid of parasites such as mites and ticks. Renewal of the skin by moulting is supposed to allow growth in some animals such as insects; however, this has been disputed in the case of snakes.

Molting occurs periodically throughout a snake's life. Before a molt, the snake stops eating and often hides or moves to a safe place. Just before shedding, the skin becomes dull and dry looking and the eyes become cloudy or blue-coloured. The inner surface of the old skin liquefies. This causes the old skin to separate from the new skin beneath it. After a few days, the eyes clear and the snake "crawls" out of its old skin. The old skin breaks near the mouth and the snake wriggles out, aided by rubbing against rough surfaces. In many cases, the cast skin peels backward over the body from head to tail in one piece, like pulling a sock off inside-out. A new, larger, brighter layer of skin has formed underneath.

An older snake may shed its skin only once or twice a year. But a younger snake, still growing, may shed up to four times a year. The discarded skin gives a perfect imprint of the scale pattern, and it is usually possible to identify the snake if the discarded skin is reasonably intact. This periodic renewal has led to the snake being a symbol of healing and medicine, as pictured in the Rod of Asclepius.

Perception

Eyesight: Snake vision varies greatly, from only being able to distinguish light from dark to keen eyesight, but the main trend is that their vision is adequate although not sharp, and allows them to track movements. Generally, vision is best in arboreal snakes and weakest in burrowing snakes. Some snakes, such as the Asian vine snake (genus *Ahaetulla*), have binocular vision, with both eyes capable of focusing on the same point. Most snakes focus by moving the lens back and forth in relation to the retina, while in the other amniote groups, the lens is stretched.

Smell: Snakes use smell to track their prey. They smell by using their forked tongues to collect airborne particles, then passing them to the vomeronasal organ or *Jacobson's organ* in the mouth for examination. The fork in the tongue gives snakes a sort of directional sense of smell and taste simultaneously. They keep their tongues constantly in motion, sampling particles from the air, ground, and water, analysing the chemicals found, and determining the presence of prey or predators in the local environment. In water-dwelling snakes, such as the Anaconda, the tongue functions efficiently under water.

Vibration sensitivity: The part of the body in direct contact with the ground is very sensitive to vibration; thus, a snake can sense other animals approaching by detecting faint vibrations in the air and on the ground.

Infrared sensitivity: Pit vipers, pythons, and some boas have infrared-sensitive receptors in deep grooves between the nostril and eye, although some have labial pits on their upper lip just below the nostrils (common in pythons), which allow them to "see" the radiated heat of warm-blooded prey mammals.

Chapter 5

Anatomy and Physiology of Animals Sensitive Organs

Cardiovascular System/Blood Circulation

The circulatory system is the continuous system of tubes through which the blood is pumped around the body. It supplies the tissues with their requirements and removes waste products. In mammals and birds the blood circulates through two separate systems - the first from the heart to the lungs and back to the heart again (the pulmonary circulation) and the second from the heart to the head and body and back again. The tubes through which the blood flows are the arteries, capillaries and veins. The heart pumps blood into arteries that carry it away from the heart. The arteries divide into very thin vessels called capillaries that form a network between the cells of the body. The capillaries then join up again to make veins that return the blood to the heart.

Arteries

Arteries carry blood away from the heart. They have thick elastic walls that stretch and can withstand the surges of high pressure blood caused by the heartbeat (the pulse). The arteries divide into smaller vessels called arterioles. The hole down the centre of the artery is called the lumen. There are three layers of tissue in the walls of an artery. It is lined with squamous epithelial cells. The middle layer is the thickest layer. It made of elastic fibres and smooth muscle to make it stretchy. The outer fibrous layer protects the artery. The pulse is only felt in arteries.

The Pulse

The pulse is the spurt of high pressure blood that passes along the aorta and arteries when the left ventricle contracts. As the pulse of blood passes along an artery the elastic walls stretch. When the pulse has passed the walls contract and this helps push the blood along. The pulse is easily felt at certain places where an artery passes near the surface of the body. It is strongest near the heart and becomes weaker as it travels away from the heart. The pulse disappears altogether in the capillaries.

Capillaries

Arterioles divide repeatedly to form a network penetrating between the cells of all tissues of the body. These small vessels are called capillaries. The walls are only one cell thick and some capillaries are so narrow that red blood cells have to fold up to pass through them. Capillaries form networks in tissues called capillary beds. The capillary networks in capillary beds are so dense that no living cell is far from its supply of oxygen and food.

Note: All arteries carry oxygenated blood except for the pulmonary artery that carries deoxygenated blood to the lungs.

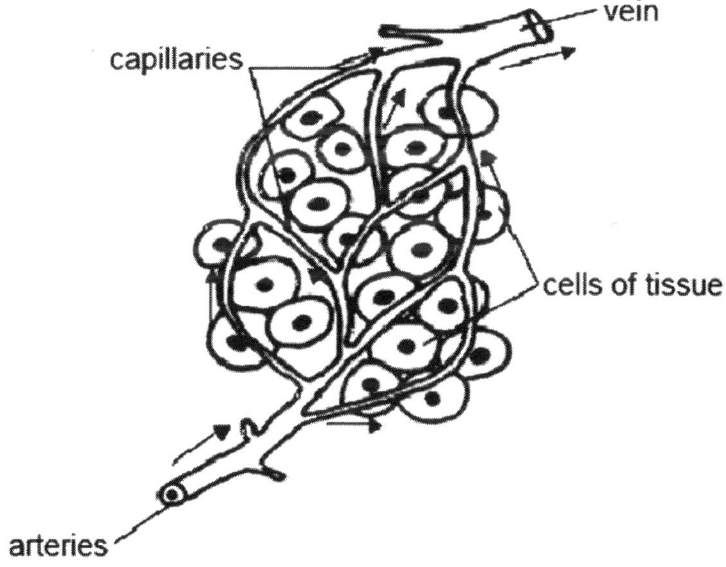

Figure: *A capillary bed*

The Formation of Tissue Fluid and Lymph

The thin walls of capillaries allow water, some white blood cells and many dissolved substances to diffuse through them. These form a clear fluid called tissue fluid (or extracellular fluid or interstitial fluid) that surrounds the cells of the tissues. The tissue fluid allows oxygen and nutrients to pass from the blood to the cells and carbon dioxide and other waste products to be removed from the tissues.

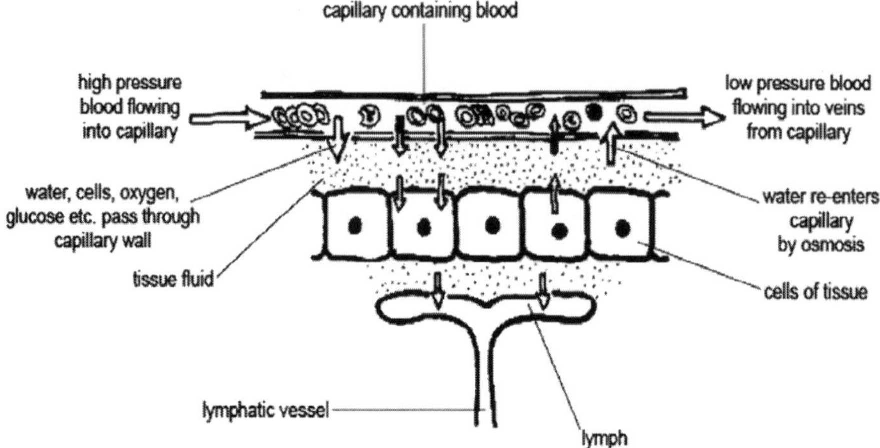

Figure 2: *The formation of tissue fluid and lymph from blood*

Some tissue fluid finds its way back into the capillaries and some of it flows into the blind-ended lymphatic vessels that form a network in the tissues. Once the tissue fluid has entered the lymphatics it is called lymph although its composition remains the same. The lymph vessels have walls that are even thinner than the capillaries. This means that molecules and particles that are larger than those that can pass into the blood stream e.g. cancer cells and bacteria can enter the lymphatic system. These are then filtered out as the lymph passes through lymph nodes.

Veins

Capillaries unite to form larger vessels called venules that join to form veins. Veins return blood to the heart and since blood that flows in veins has already passed through the fine capillaries, it flows slowly with no pulse and at low pressure. For this reason veins have thinner walls than arteries although there have the same three layers in them as arteries. As there is no pulse in veins, the blood is squeezed along them by the contraction of the skeletal muscles that lay alongside them. Veins also have valves in them that prevent blood flowing backwards.

The pulmonary vein that carries oxygenated blood from the lungs to the left atrium of the heart is an exception.

Regulation of Blood Flow

The flow of blood along arteries, arterioles and capillaries is not constant but can be controlled depending upon the requirements of the body. For example more blood is directed to the skeletal muscles, brain or digestive system when they are active. Regulation of the blood flow to the arterioles of the skin is also important in controlling body temperature. The size of the vessels is adjusted by the contraction or relaxation of smooth muscle fibres in their walls.

Oedema and Fluid Loss

Oedema is the swelling of the tissues due to the accumulation of tissue fluid. This may occur because the tissue fluid is prevented from returning to the bloodstream and accumulates in the tissues. This may be caused by physical inactivity (e.g. long car or plane trips in humans) or because of imbalances in the proteins in the blood. This is what causes the "potbelly" of the malnourished child or worm-infested puppy.

Loss of body fluid can be caused not only by drinking insufficient liquid but also through diarrhea and vomiting or sudden loss of blood due to haemorrhage. The effect is to reduce the volume of the blood which decreases the blood pressure. This could be dangerous because the supply of adequate blood to the brain depends upon maintaining the blood pressure at a constant level.

To compensate for the loss of fluid various mechanisms come into play. First of all the blood vessels contract in order to try and maintain the pressure. Then, since the loss of fluid tends to make the blood more concentrated and increases its osmotic pressure, fluid is drawn into the blood from the tissues by osmosis.

The Spleen

The spleen is situated near the stomach. It has a rich blood supply and acts as a reservoir of red blood cells. When there is a sudden loss of blood, as happens when a haemorrhage occurs, the spleen contracts to release large numbers of red blood cells into the circulation. The spleen also destroys old red cells and makes new lymphocytes but it is not an essential organ because its removal in adult life seems to cause few problems. In the foetus, the spleen makes both red and white cells.

Important Blood Vessels of The Systemic (Body) Circulation

Blood is pumped out into the body via the main artery, the aorta. This takes the blood to the head, the limbs and all the body organs. After passing through a network of fine capillaries, the blood is returned to the heart in the largest vein, the vena cava.

Arteries and veins to and from many organs often run alongside each other and have the same name e.g. the renal artery and vein serve the kidney, the femoral artery and vein serve the hind limbs and the subclavian artery and vein serve the forelimbs. However, blood to the head passes along the carotid artery and returns to the cranial vena cava via the jugular vein.

One variation on this arrangement is found in the blood vessels that serve the digestive tract. A variety of arteries take blood from the aorta to the intestines but blood from the intestines is carried by the hepatic portal vein to the liver where the digested food can be processed. This vessel is unlike others in that it transports blood from one organ to another rather than to or from the heart like arteries or veins.

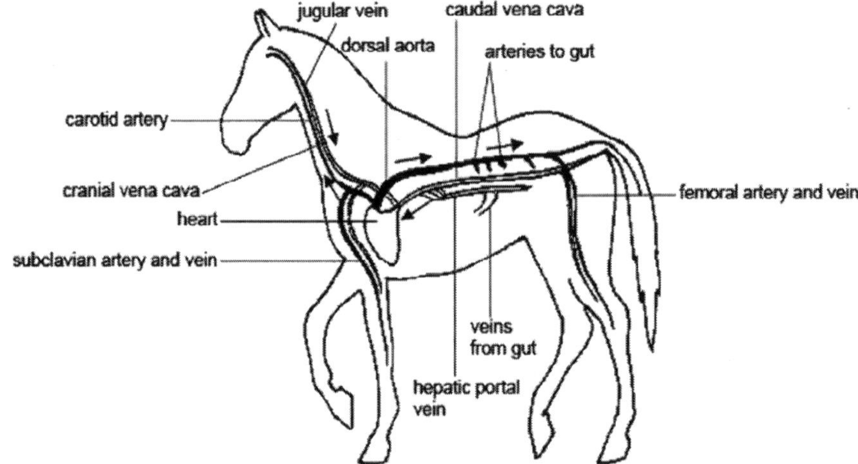

Figure: *The main arteries and veins of the horse*

Blood Pressure

The blood pressure is the pressure of the blood against the walls of the main arteries. The pressure is highest as the pulse produced by the contraction of the left ventricle passes along the artery. This is known as the systolic pressure. Pressure is much lower between pulses. This is known as the diastolic pressure. Blood pressure is measured in millimetres of mercury. A blood pressure that is higher than expected is known as hypertension while a pressure lower than expected is known as hypotension.

- The circulatory system is double with the blood passing through the heart twice.
- Arteries carry blood away from the heart. They have thick elastic walls that stretch and can withstand the high pressure of the pulse.
- Capillaries are small, thin walled vessels that form a network between the cells of the tissues.

- Veins return low pressure blood to the heart. They have thinner walls than arteries.
- The pulse is the spurt of high pressure blood that passes along the arteries when the left ventricle contracts. It can be felt where arteries pass close to the body surface.
- Tissue fluid is the clear fluid that leaks from the capillaries and surrounds the cells of the tissues. Lymph forms when tissue fluid enters lymphatics.
- Important blood vessels include the vena cava, aorta, pulmonary artery, carotid artery, jugular vein, renal artery and vein and hepatic portal vessel.

Respiratory System

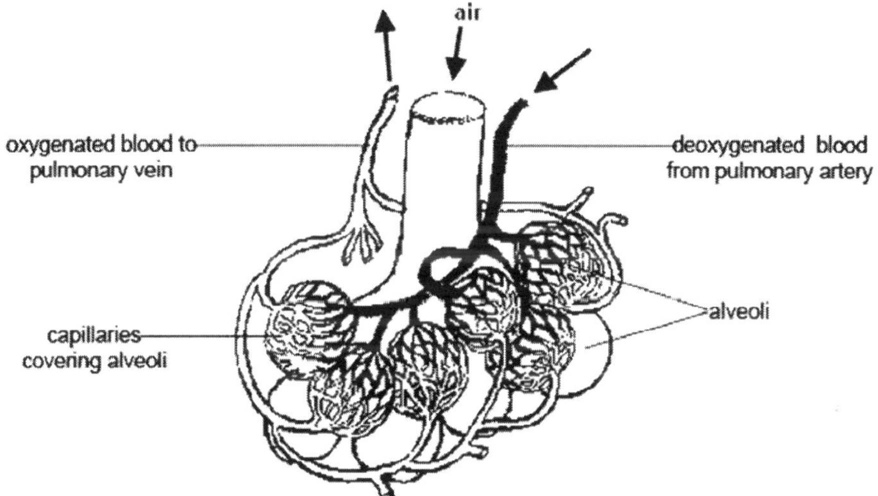

Figure: *Alveoli with blood supply*

Animals require a supply of energy to survive. This energy is needed to build large molecules like proteins and glycogen, make the structures in cells, move chemicals through membranes and around cells, contract muscles, transmit nerve impulses and keep the body warm. Animals get their energy from the large molecules that they eat as food. Glucose is often the energy source but it may also come from other carbohydrates, as well as fats and protein. The energy is made by the biochemical process known as cellular respiration that takes place in the mitochondria inside every living cell.

The overall reaction can be summarised by the word equation given below.

Charbohydrate Food (glucose) + Oxygen = Carbon Dioxide + Water + energy

As you can see from this equation, the cells need to be supplied with oxygen and glucose and the waste product, carbon dioxide, which is poisonous to cells, needs to be removed. The way the digestive system provides the glucose for cellular respiration will be described in Chapter 5, but here we are only concerned with the two gases, oxygen and carbon dioxide, that are involved in cellular respiration. These gases are carried in the blood to and from the tissues where they are required or produced. Oxygen enters the body from the air (or water in fish)and carbon dioxide is usually eliminated from the same part of the body. This process is called gas exchange. In fish gas exchange occurs in the gills, in land dwelling vertebrates lungs are the gas exchange organs and frogs use gills when they are tadpoles and lungs, the mouth and the skin when adults.

Mammals (and birds) are active and have relatively high body temperatures so they require large amounts of oxygen to provide sufficient energy through cellular respiration. In order to take in enough oxygen and release all the carbon dioxide produced they need a very large surface area over which gas exchange can take place. The many minute air sacs or alveoli of the lungs provide this. When you look at these under the microscope they appear rather like bunches of grapes covered with a dense network of fine capillaries. A thin layer of water covers the inner surface of each alveolus. There is only a very small distance -just 2 layers of thin cells - between the air in the alveoli and the blood in the capillaries. The gases pass across this gap by diffusion.

Diffusion And Transport of Oxygen

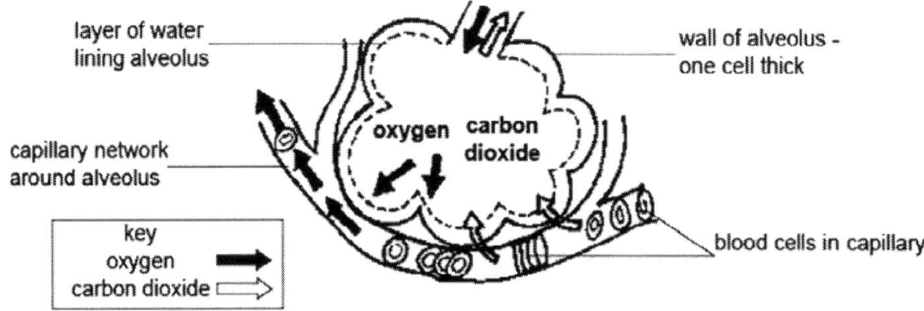

Figure: *Cross section of an alveolus*

The air in the alveoli is rich in oxygen while the blood in the capillaries around the alveoli is deoxygenated. This is because the haemoglobin in the red blood cells has released all the oxygen it has been carrying to the cells of the body. Oxygen diffuses from high concentration to low concentration. It therefore crosses the narrow barrier between the alveoli and the capillaries to enter the blood and combine with the haemoglobin in the red blood cells to form oxyhaemoglobin.

The narrow diametre of the capillaries around the alveoli means that the blood flow is slowed down and that the red cells are squeezed against the capillary walls. Both of these factors help the oxygen diffuse into the blood.

When the blood reaches the capillaries of the tissues the oxygen splits from the haemoglobin molecule. It then diffuses into the tissue fluid and then into the cells.

Diffusion and Transport of Carbon Dioxide

Blood entering the lung capillaries is full of carbon dioxide that it has collected from the tissues. Most of the carbon dioxide is dissolved in the plasma either in the form of sodium bicarbonate or carbonic acid. A little is transported by the red blood cells. As the blood enters the lungs the carbon dioxide gas diffuses through the capillary and alveoli walls into the water film and then into the alveoli. Finally it is removed from the lungs during breathing out.

The Air Passages

When air is breathed in it passes from the nose to the alveoli of the lungs down a series of tubes. After entering the nose the air passes through the nasal cavity, which is lined with a moist membrane that adds warmth and moisture to the air as it passes. The air then flows through the pharynx or throat, a passage that carries both food and air, to the larynx where the voice-box is located. Here the passages for food and air separate again. Food must pass into the oesophagus and the air into the windpipe or trachea. To prevent food entering this, a small flap of tissue called the epiglottis closes the opening during swallowing. A reflex that inhibits breathing during swallowing also (usually) prevents choking on food.

The trachea is the tube that ducts the air down the throat. Incomplete rings of cartilage in its walls help keep it open even when the neck is bent and head turned. The fact that acrobats and people that tie themselves in knots doing yoga still keep breathing during the most contorted manoeuvres shows how effective this arrangement is. The air passage now divides into the two bronchi that take the air to the right and left lungs before dividing into smaller and smaller bronchioles that spread throughout the lungs to carry air to the alveoli. Smooth muscles in the walls of the bronchi and bronchioles adjust the diametre of the air passages.

The tissue lining the respiratory passages produces mucus and is covered with minute hairs or cilia. Any dust that is breathed into the respiratory system immediately gets entangled in the mucous and the cilia move it towards the mouth or nose where it can be coughed up or blown out.

The Lungs and the Pleural Cavities

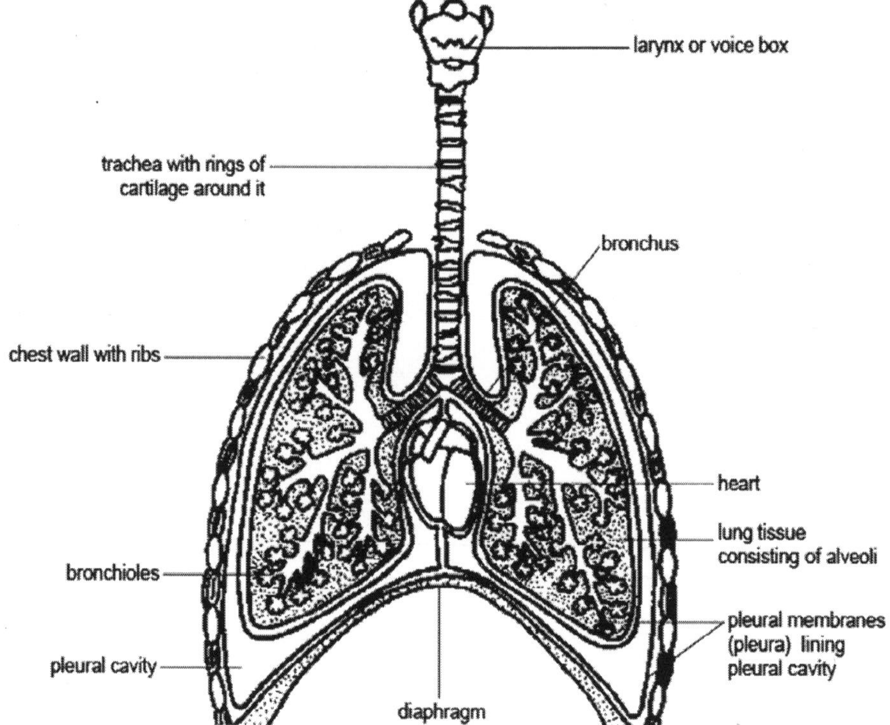

Figure: *The respiratory system*

The lungs fill most of the chest or thoracic cavity, which is completely separated from the abdominal cavity by the diaphragm. The lungs and the spaces in which they lie (called the pleural cavities) are covered with membranes called the pleura. There is a thin film of fluid between the two membranes. This lubricates them as they move over each other during breathing movements.

Collapsed Lungs

The pleural cavities are completely airtight with no connection with the outside and if they are punctured by accident (a broken rib will often do this), air rushes in and the lung collapses. Separating the two lungs is a region of tissue that contains the oesophagus, trachea, aorta, vena cava and lymph nodes. This is called the mediastinum. In humans and sheep it separates the cavity completely so that puncturing one pleural cavity leads to the collapse of only one lung. In dogs, however, this separation is incomplete so a puncture results in a complete collapse of both lungs.

Breathing

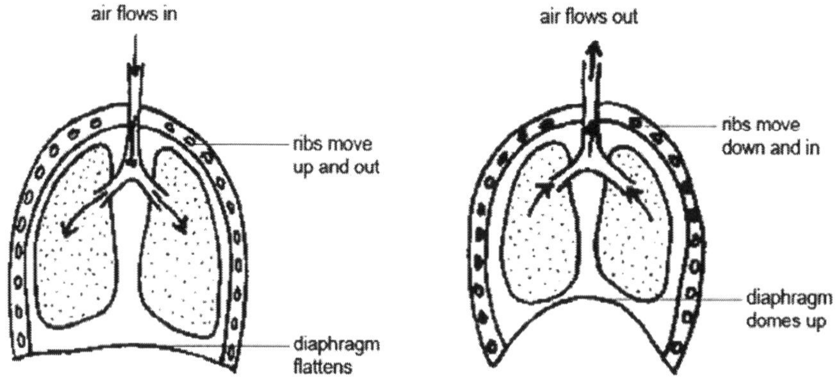

Figure: *(a) Inspiration; (b) Expiration*

The process of breathing moves air in and out of the lungs. Sometimes this process is called respiration but it is important not to confuse it with the chemical process, cellular respiration, that takes place in the mitochondria of cells. Breathing is brought about by the movement of the diaphragm and the ribs.

Inspiration

The diaphragm is a thin sheet of muscle that completely separates the abdominal and thoracic cavities. When at rest it domes up into the thoracic cavity but during breathing in or inspiration it flattens. At the same time special muscles in the chest wall move the ribs forwards and outwards. These movements of both the diaphragm and the ribs cause the volume of the thorax to increase. Because the pleural cavities are airtight, the lungs expand to fill this increased space and air is drawn down the trachea into the lungs.

Expiration

Expiration or breathing out consists of the opposite movements. The ribs move down and in and the diaphragm resumes its domed shape so the air is expelled. Expiration is usually passive and no energy is required (unless you are blowing up a balloon).

Lung Volumes

As you sit here reading this just pay attention to your breathing. Notice that your in and out breaths are really quite small and gentle (unless you have just rushed here from somewhere else!). Only a small amount of the total volume that your lungs hold is breathed in and out with each breath.

This kind of gentle "at rest" breathing is called tidal breathing and the volume breathed in or out (they should be the same) is the tidal volume. Sometimes people want to measure the volume of air inspired or expired during a minute of this normal breathing. This is called the minute volume.

It could be estimated by measuring the volume of one tidal breath and then multiplying that by the number of breaths in a minute. Of course it is possible to take a deep breath and breathe in as far as you can and then expire as far as possible. The volume of the air expired when a maximum expiration follows a maximum inspiration is called the vital capacity.

Composition of Air

The air animals breathe in consists of 21% oxygen and 0.04% carbon dioxide. Expelled air consists of 16% oxygen and 4.4% carbon dioxide. This means that the lungs remove only a quarter of the oxygen contained in the air. This is why it is possible to give someone (or an animal) artificial respiration by blowing expired air into their mouth.

Breathing is usually an unconscious activity that takes place whether you are awake or asleep, although, humans at least, can also control it consciously. Two regions in the hindbrain called the medulla oblongata and pons control the rate of breathing. These are called respiratory centres. They respond to the concentration of carbon dioxide in the blood. When this concentration rises during a bout of activity, for example, nerve impulses are automatically sent to the diaphragm and rib muscles that increase the rate and the depth of breathing. Increasing the rate of breathing also increases the amount of oxygen in the blood to meet the needs of this increased activity.

The Acidity of the Blood and Breathing

The degree of acidity of the blood (the acid-base balance) is critical for normal functioning of cells and the body as a whole. For example, blood that is too acidic or alkaline can seriously affect nerve function causing a coma, muscle spasms, convulsions and even death. Carbon dioxide carried in the blood makes the blood acidic and the higher the concentration of carbon dioxide the more acidic it is. This is obviously dangerous so there are various mechanisms in the body that bring the acid-base balance back within the normal range. Breathing is one of these homeostatic mechanisms. By increasing the rate of breathing the animal increases the amount of dissolved carbon dioxide that is expelled from the blood. This reduces the acidity of the blood.

Breathing In Birds

Birds have a unique respiratory system that enables them to respire at the very high rates necessary for flight. The lungs are relatively solid structures that do not change shape and size in the same way as mammalian lungs do. Tubes run through them and connect with a series of air sacs situated in the thoracic and abdominal body cavities and some of the bones. Movements of the ribs and breastbone or sternum expand and compress these air sacs so they act rather like bellows and pump air through the lungs. The evolution of this extremely efficient system of breathing has enabled birds to migrate vast distances and fly at altitudes higher than the summit of Everest.

- Animals need to breathe to supply the cells with oxygen and remove the waste product carbon dioxide.
- The lungs are situated in the pleural cavities of the thorax.
- Gas exchange occurs in the alveoli of the lungs that provide a large surface area. Here oxygen diffuses from the alveoli into the red blood cells in the capillaries that surround the alveoli. Carbon dioxide, at high concentration in the blood, diffuses into the alveoli to be breathed out.
- Inspiration occurs when muscle contraction causes the ribs to move up and out and the diaphragm to flatten. These movements increase the volume of the pleural cavity and draw air down the respiratory system into the lungs.
- The air enters the nasal cavity and passes to the pharynx and larynx where the epiglottis closes the opening to the lungs during swallowing. The air passes down the trachea kept open by rings of cartilage to the bronchi and bronchioles and then to the alveoli.
- Expiration is a passive process requiring no energy as it relies on the relaxation of the muscles and recoil of the elastic tissue of the lungs.
- The rate of breathing is determined by the concentration of carbon dioxide in the blood. As carbon dioxide makes blood acidic, the rate of breathing helps control the acid/base balance of the blood.

- The cells lining the respiratory passages produce mucus which traps dust particles, which are wafted into the nose by cilia.

Lymphatic System

When tissue fluid enters the small blind-ended lymphatic capillaries that form a network between the cells it becomes lymph. Lymph is a clear watery fluid that is very similar to blood plasma except that it contains large numbers of white blood cells, mostly lymphocytes. It also contains protein, cellular debris, foreign particles and bacteria. Lymph that comes from the intestines also contains many fat globules following the absorption of fat from the digested food into the lymphatics (lacteals) of the villi. From the lymph capillaries the lymph flows into larger tubes called lymphatic vessels. These carry the lymph back to join the blood circulation.

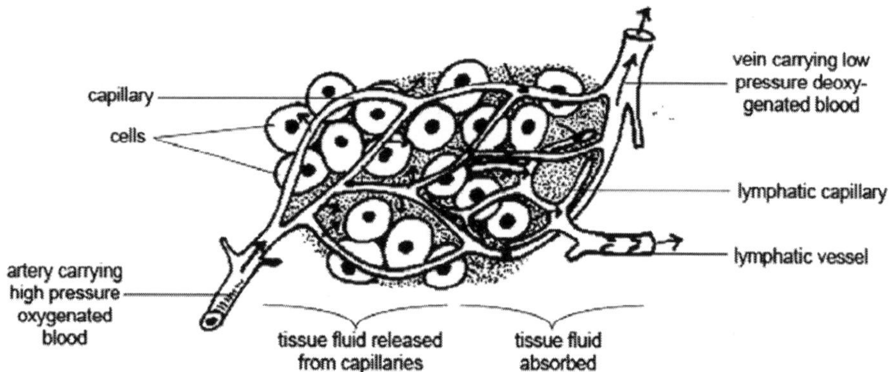

Figure: *A capillary bed with lymphatic capillaries*

Lymphatic Vessels

Lymphatic vessels have several similarities to veins. Both are thin walled and return fluid to the right hand side of the heart. The movement of the fluid in both is brought about by the contraction of the muscles that surround them and both have valves to prevent backflow. One important difference is that lymph passes through at least one lymph node or gland before it reaches the blood system. These filter out used cell parts, cancer cells and bacteria and help defend the body from infection. Lymph nodes are of various sizes and shapes and found throughout the body and the more important ones are shown. They consist of lymph tissue surrounded by a fibrous sheath. Lymph flows into them through a number of incoming vessels. It then trickles through small channels where white cells called macrophages (derived from monocytes) remove the bacteria and debris by engulfing and digesting them. The lymph then leaves the lymph nodes through outgoing vessels to continue its journey towards the heart where it rejoins the blood circulation.

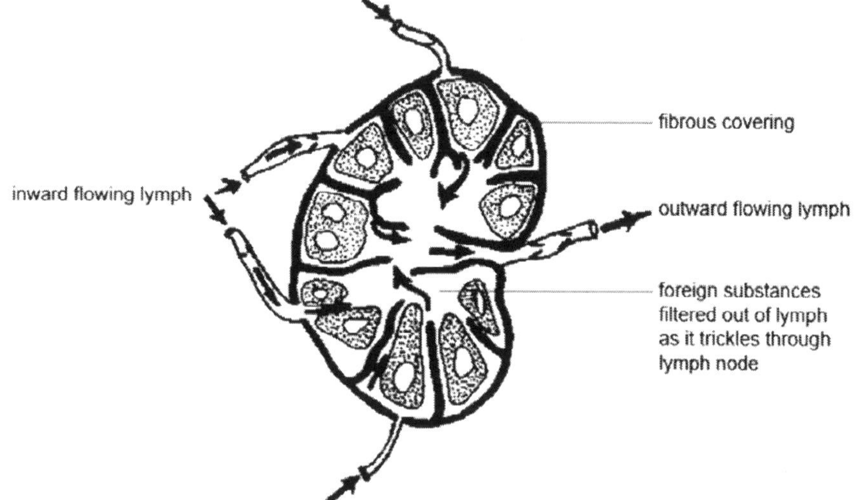

Figure: *A lymph node*

As well as filtering the lymph, lymph nodes produce the white cells known as lymphocytes. Lymphocytes are also produced by the thymus, spleen and bone marrow. There are two kinds of lymphocyte. The first attach invading micro organisms directly while others produce antibodies that circulate in the blood and attack them.

The function of the lymphatic system can therefore be summarised as transport and defence. It is important for returning the fluid and proteins that have escaped from the blood capillaries to the blood system and is also responsible for picking up the products of fat digestion in the small intestine. Its other essential function is as part of the immune system, defending the body against infection.

Problems with Lymph Nodes and the Lymphatic System

During infection of the body the lymph nodes often become swollen and tender because of their increased activity. This is what causes the swollen 'glands' in your neck during throat infections, mumps and tonsillitis. Sometimes the bacteria multiply in the lymph node and cause inflammation. Cancer cells may also be carried to the lymph nodes and then transported to other parts of the body where they may multiply to form a secondary growth or metastasis. The lymphatic system may therefore contribute to the spread of cancer. Inactivity of the muscles surrounding the lymphatic vessels or blockage of these vessels causes tissue fluid to 'back up' in the tissues resulting in swelling or oedema.

Other Organs of the Lymphatic System

The spleen is an important part of the lymphatic system. It is a deep red organ situated in the abdomen caudal to the stomach. It is composed of two different types of tissue. The first type makes and stores lymphocytes, the cells of the immune system. The second type of tissue destroys worn out red blood cells, breaking down the haemoglobin into iron, which is recycled, and waste products that are excreted. The spleen also stores red blood cells. When severe blood loss occurs, it contracts and releases them into the circulation.

The thymus is a large pink organ lying just under the sternum (breastbone) just cranial to the heart. It has an important function processing lymphocytes so they are capable of recognising and attacking foreign invaders like bacteria. Other lymph organs are the bone marrow of the long bones where lymphocytes are produced and lymph nodules, which are like tiny lymph nodes. Large clusters of these are found in the wall of the small intestine (called Peyer's Patches) and in the tonsils.

The Gut and Digestion

Plant cells are made of organic molecules using energy from the sun. This process is called photosynthesis. Animals rely on these ready-made organic molecules to supply them with their food. Some animals (herbivores) eat plants; some (carnivores) eat the herbivores.

Herbivores

Although herbivores eat plant material they have a basic problem. Their own digestive enzymes do not break down the large cellulose molecules in the plant cell walls. Micro-organisms like bacteria, on the other hand, can break them down. Therefore herbivores employ micro-organisms to do the job for them.

There are two types of herbivore:

- The first, ruminants like cattle, sheep and goats, house these bacteria in a special compartment in the enlarged stomach called the rumen.
- The second group has an enlarged large intestine and caecum, called a functional caecum, occupied by cellulose digesting micro-organisms. These non-ruminant herbivores include the horse, rabbit and rat.

Even with the help of microorganisms, eating plants still poses problems. Plants are not a good source of nutrients, and herbivores have to eat large quantities of food to obtain all they require. Herbivores like cows, horses and rabbits typically spend much of their day feeding. To give the micro- organisms access to the cellulose molecules, the plant cell walls need to be broken down. This is why herbivores have teeth that are adapted to crush and grind. Their guts also tend to be lengthy and the food takes a long time to pass through it. Eating plants does have some advantages. Plants are immobile so herbivores normally have to spend little energy collecting them. This contrasts with the other main group of animals- the carnivores that often have to chase their prey.

Carnivores

Carnivorous animals like those in the cat and dog families, bears, seals, crocodiles and birds of prey catch and eat other animals. They often have to use large amounts of energy finding, stalking, catching and killing their prey.

However, they are rewarded by the fact that meat provides a very concentrated source of nutrients. Carnivores in the wild therefore tend to eat distinct meals often with long and irregular intervals between them. Time after feeding is spent digesting and absorbing the food. The guts of carnivores are usually shorter and less complex than those of herbivores because meat is easier to digest than plant material. Carnivores usually have teeth that are specialised for dealing with flesh, gristle and bone. They have sleek bodies, strong, sharp claws and keen senses of smell, hearing and sight. They are also often cunning, alert and have an aggressive nature.

Omnivores

Many animals feed on both animal and vegetable material – they are omnivorous. Most primates including humans belong to this category as do pigs and rats. Their food is diverse, ranging from plant material to animals they have either killed themselves or scavenged from other carnivores. Omnivores lack the specialised teeth and guts of carnivores and herbivores but are often highly intelligent and adaptable reflecting their varied diet.

Treatment of Food

Whether an animal eats plants or flesh, the carbohydrates, fats and proteins in the food it eats are generally giant molecules. These need to be split up into smaller ones before they can pass into the blood and enter the cells to be used for energy or to make new cell constituents.

For example:

- Carbohydrates like cellulose, starch, and glycogen need to be split into glucose and other monosaccharides;
- Proteins need to be split into amino acids;
- Fats or lipids need to be split into fatty acids and glycerol.

The Gut

The digestive tract, alimentary canal or gut is a hollow tube stretching from the mouth to the anus. It is the organ system concerned with the treatment of foods. At the mouth the large food molecules are taken into the gut - this is called ingestion. They must then be broken down into smaller ones by digestive enzymes - digestion, before they can be taken from the gut into the blood stream - absorption. The cells of the body can then use these small molecules - assimilation. The indigestible waste products are eliminated from the body by the act of egestion.

The 4 major functions of the gut are:

1. Transporting the food;
2. Processing the food physically by breaking it up (chewing), mixing, adding fluid etc.
3. Processing the food chemically by adding digestive enzymes to split large food molecules into smaller ones.
4. Absorbing these small molecules into the blood stream so the body can use them.

The regions of a typical mammals gut (for example a cat or dog) are shown.

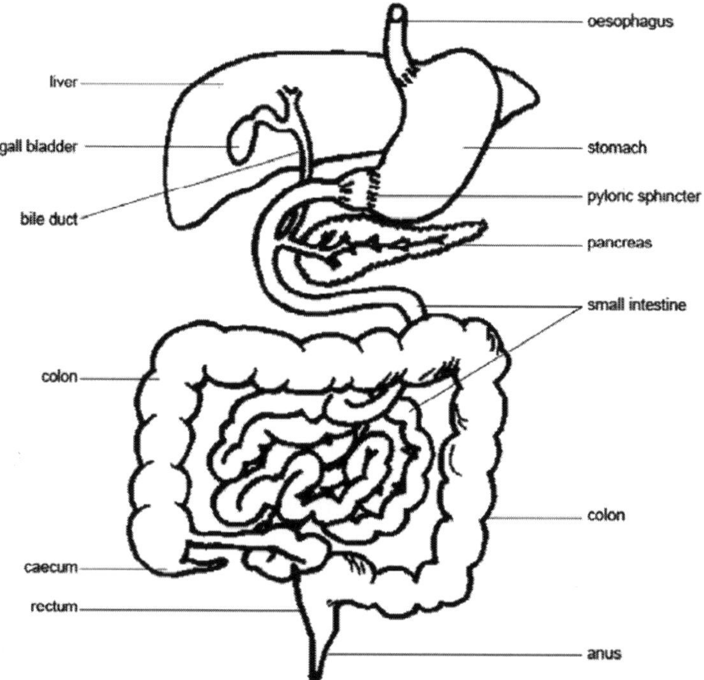

***Figure:** A typical mammalian gut*

The food that enters the mouth passes to the oesophagus, then to the stomach, small intestine, caecum, large intestine, rectum and finally undigested material exits at the anus. The liver and pancreas produce secretions that aid digestion and the gall bladder stores bile.

Mouth

The mouth takes food into the body. The lips hold the food inside the mouth during chewing and allow the baby animal to suck on its mother's teat. In elephants the lips (and nose) have developed into the trunk which is the main food collecting tool. Some mammals, e.g. hamsters, have stretchy cheek pouches that they use to carry food or material to make their nests. The sight or smell of food and its presence in the mouth stimulates the salivary glands to secrete saliva. There are four pairs of these glands in cats and dogs. The fluid they produce moistens and softens the food making it easier to swallow. It also contains the enzyme, salivary amylase, which starts the digestion of starch.

The tongue moves food around the mouth and rolls it into a ball for swallowing. Taste buds are located on the tongue and in dogs and cats it is covered with spiny projections used for grooming and lapping. The cow's tongue is prehensile and wraps around grass to graze it. Swallowing is a complex reflex involving 25 different muscles. It pushes food into the oesophagus and at the same time a small flap of tissue called the epiglottis closes off the windpipe so food doesn't go 'down the wrong way' and choke the animal.

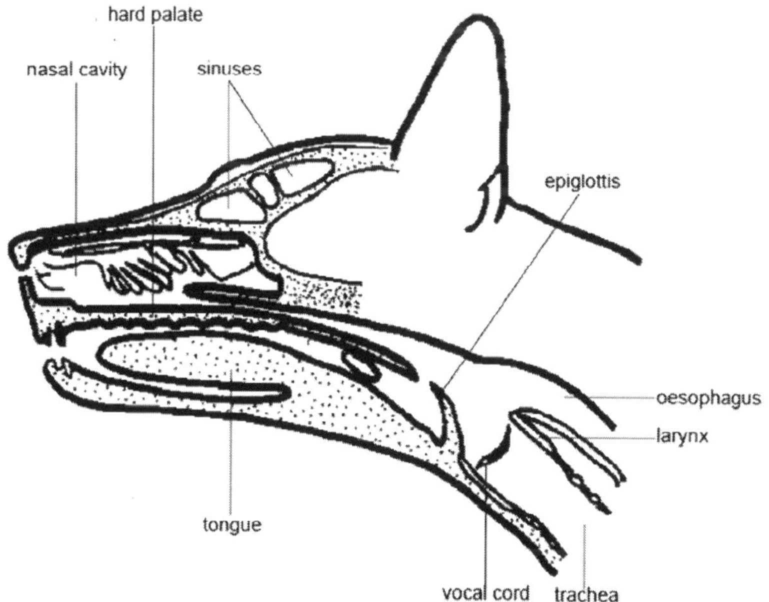

Figure: *Section through the head of a dog*

Teeth

Teeth seize, tear and grind food. They are inserted into sockets in the bone and consist of a crown above the gum and root below. The crown is covered with a layer of enamel, the hardest substance in the body. Below this is the dentine, a softer but tough and shock resistant material. At the centre of the tooth is a space filled with pulp which contains blood vessels and nerves. The tooth is cemented into the socket and in most teeth the tip of the root is quite narrow with a small opening for the blood vessels and nerves.

In teeth that grow continuously, like the incisors of rodents, the opening remains large and these teeth are called open rooted teeth. Mammals have 2 distinct sets of teeth. The first the milk teeth are replaced by the permanent teeth.

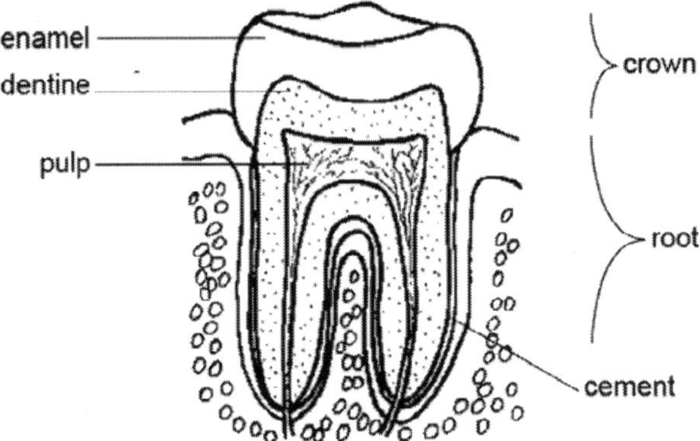

***Figure:** Structure of a tooth*

Types of Teeth

All the teeth of fish and reptiles are similar but mammals usually have four different types of teeth.

The incisors are the chisel-shaped 'biting off' teeth at the front of the mouth. In rodents and rabbits the incisors never stop growing (open-rooted teeth). They must be worn or ground down continuously by gnawing. They have hard enamel on one surface only so they wear unevenly and maintain their sharp cutting edge.

The largest incisors in the animal kingdom are found in elephants, for tusks are actually giant incisors. Sloths have no incisors at all, and sheep have no incisors in the upper jaw. Instead there is a horny pad against which the bottom incisors cut. The canines or 'wolf-teeth' are long, cone-shaped teeth situated just behind the incisors. They are particularly well developed in the dog and cat families where they are used to hold, stab and kill the prey.

The tusks of boars and walruses are large canines while rodents and herbivores like sheep have no (or reduced) canines. In these animals the space where the canines would normally be is called the diastema. In rodents like the rat and beaver it allows the debris from gnawing to be expelled easily.

The cheek teeth or premolars and molars crush and grind the food. They are particularly well developed in herbivores where they have complex ridges that form broad grinding surfaces. These are created from alternating bands of hard enamel and softer dentine that wear at different rates.

In carnivores the premolars and molars slice against each other like scissors and are called carnassial teeth. They are used for shearing flesh and bone.

Dental Formula

The numbers of the different kinds of teeth can be expressed in a dental formula. This gives the numbers of incisors, canines, premolars and molars in one half of the mouth. The numbers of these four types of teeth in the left or right half of the upper jaw are written above a horizontal line and the four types of teeth in the right or left half of the lower jaw are written below it.

It indicates that in the upper right (or left) half of the jaw there are no incisors or canines (i.e. there is a diastema), three premolars and three molars. In the lower right (or left) half of the jaw are three incisors, one canine, three premolars and three molars.

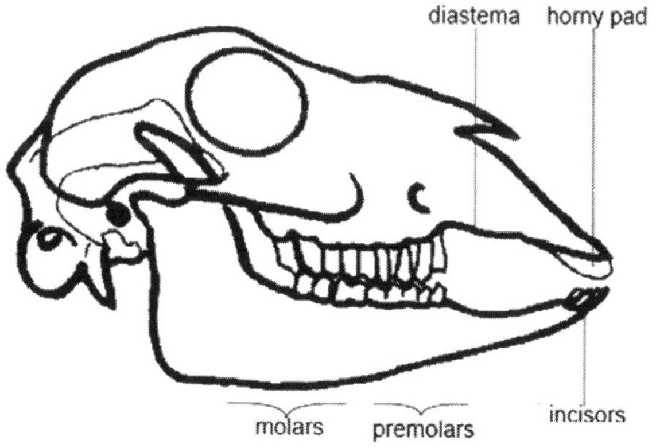

Figure: *A sheep's skull*

The formula indicates that in the right (or left) half of the upper jaw there are three incisors, one canine, four premolars and two molars. In the right (or left) half of the lower jaw there are three incisors, one canine, four premolars and three molars.

Oesophagus

The oesophagus transports food to the stomach. Food is moved along the oesophagus, as it is along the small and large intestines, by contraction of the smooth muscles in the walls that push the food along rather like toothpaste along a tube. This movement is called peristalsis.

Stomach

The stomach stores and mixes the food. Glands in the wall secrete gastric juice that contains enzymes to digest protein and fats as well as hydrochloric acid to make the contents very acidic. The walls of the stomach are very muscular and churn and mix the food with the gastric juice to form a watery mixture called chyme (pronounced kime). Rings of muscle called sphincters at the entrance and exit to the stomach control the movement of food into and out of it.

Small Intestine

Most of the breakdown of the large food molecules and absorption of the smaller molecules take place in the long and narrow small intestine. The total length varies but it is about 6.5 metres in humans, 21 metres in the horse, 40 metres in the ox and over 150 metres in the blue whale.

It is divided into 3 sections: the duodenum (after the stomach), jejunum and ileum. The duodenum receives 3 different secretions:

1) Bile from the liver;
2) Pancreatic juice from the pancreas and
3) Intestinal juice from glands in the intestinal wall.

These complete the digestion of starch, fats and protein. The products of digestion are absorbed into the blood and lymphatic system through the wall of the intestine, which is lined with tiny finger-like projections called villi that increase the surface area for more efficient absorption.

The Rumen

In ruminant herbivores like cows, sheep and antelopes the stomach is highly modified to act as a "fermentation vat". It is divided into four parts. The largest part is called the rumen. In the cow it occupies the entire left half of the abdominal cavity and can hold up to 270 litres. The reticulum is much smaller and has a honeycomb of raised folds on its inner surface. In the camel the reticulum is further modified to store water. The next part is called the omasum with a folded inner surface. Camels have no omasum. The final compartment is called the abomasum. This is the 'true' stomach where muscular walls churn the food and gastric juice is secreted.

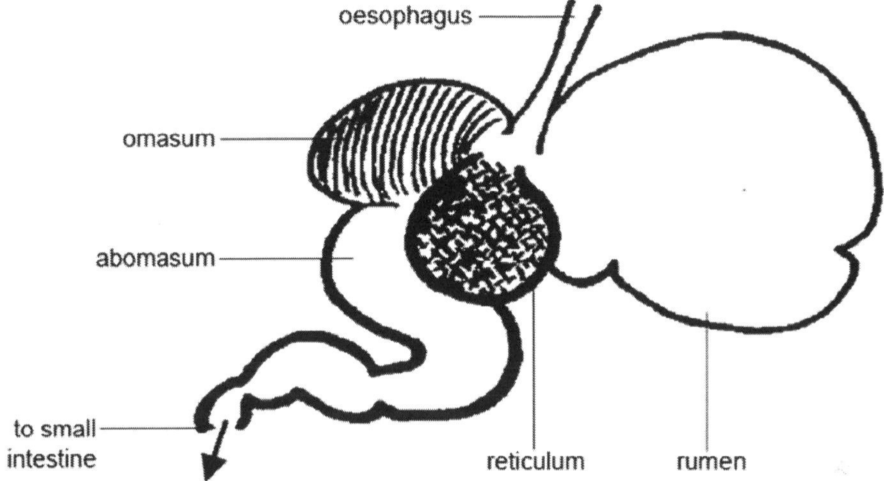

Figure: *The rumen*

Ruminants swallow the grass they graze almost without chewing and it passes down the oesophagus to the rumen and reticulum. Here liquid is added and the muscular walls churn the food. These chambers provide the main fermentation vat of the ruminant stomach. Here bacteria and single-celled animals start to act on the cellulose plant cell walls. These organisms break down the cellulose to smaller molecules that are absorbed to provide the cow or sheep with energy. In the process, the gases methane and carbon dioxide are produced. These cause the "burps" you may hear cows and sheep making.

Not only do the micro-organisms break down the cellulose but they also produce the vitamins E, B and K for use by the animal. Their digested bodies provide the ruminant with the majority of its protein requirements.

In the wild grazing is a dangerous activity as it exposes the herbivore to predators. They crop the grass as quickly as possible and then when the animal is in a safer place the food in the rumen can be regurgitated to be chewed at the animal's leisure. This is 'chewing the cud' or rumination. The finely ground food may be returned to the rumen for further work by the microorganisms or, if the particles are small enough, it will pass down a special groove in the wall of the oesophagus straight into the omasum. Here the contents are kneaded and water is absorbed before they pass to the abomasum. The abomasum acts as a "proper" stomach and gastric juice is secreted to digest the protein.

Large Intestine

The large intestine consists of the caecum, colon and rectum. The chyme from the small intestine that enters the colon consists mainly of water and undigested material such as cellulose (fibre or roughage). In omnivores like the pig and humans the main function of the colon is absorption of water to give solid faeces. Bacteria in this part of the gut produce vitamins B and K.

The caecum, which forms a dead-end pouch where the small intestine joins the large intestine, is small in pigs and humans and helps water absorption. However, in rabbits, rodents and horses, the caecum is very large and called the functional caecum. It is here that cellulose is digested by micro-organisms. The appendix, a narrow dead end tube at the end of the caecum, is particularly large in primates but seems to have no digestive function.

Functional Caecum

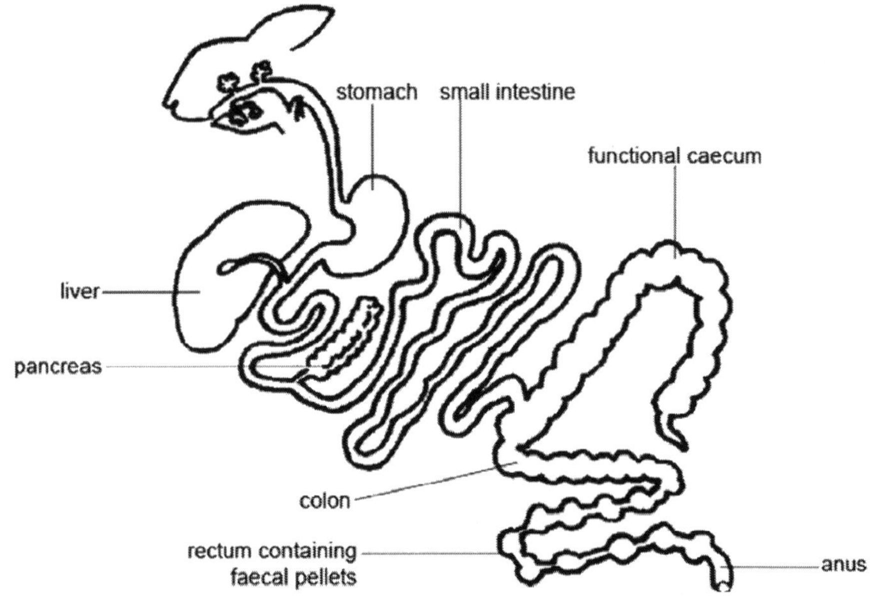

Figure: *The gut of a rabbit*

The caecum in the rabbit, rat and guinea pig is greatly enlarged to provide a "fermentation vat" for micro-organisms to break down the cellulose plant cell walls. This is called a functional caecum. In the horse both the caecum and the colon are enlarged. As in the rumen, the large cellulose molecules are broken down to smaller molecules that can be absorbed. However, the position of the functional caecum after the main areas of digestion and absorption, means it is potentially less effective than the rumen. This means that the small molecules that are produced there can not be absorbed by the gut but pass out in the faeces. The rabbit and rodents (and foals) solve this problem by eating their own faeces so that they pass through the gut a second time and the products of cellulose digestion can be absorbed in the small intestine. Rabbits produce two kinds of faeces. Softer night-time faeces are eaten directly from the anus and the harder pellets you are probably familiar with, that have passed through the gut twice.

The Gut of Birds

Birds' guts have important differences from mammals' guts. Most obviously, birds have a beak instead of teeth. Beaks are much lighter than teeth and are an adaptation for flight. Imagine a bird trying to take off and fly with a whole set of teeth in its head! At the base of the oesophagus birds have a bag-like structure called a crop. In many birds the crop stores food before it enters the stomach, while in pigeons and doves glands in the crop secretes a special fluid called crop-milk which parent birds regurgitate to feed their young. The stomach is also modified and consists of two compartments. The first is the true stomach with muscular walls and enzyme secreting glands. The second compartment is the gizzard. In seed eating birds this has very muscular walls and contains pebbles swallowed by the bird to help grind the food. This is the reason why you must always supply a caged bird with grit. In birds of prey like the falcon the walls of the gizzard are much thinner and expand to accommodate large meals.

Digestion

During digestion the large food molecules are broken down into smaller molecules by enzymes. The three most important groups of enzymes secreted into the gut are:

1. Amylases that split carbohydrates like starch and glycogen into monosaccharides like glucose.
2. Proteases that split proteins into amino acids.
3. Lipases that split lipids or fats into fatty acids and glycerol.

Glands produce various secretions which mix with the food as it passes along the gut.

These secretions include:

1. Saliva secreted into the mouth from several pairs of salivary glands. Saliva consists mainly of water but contains salts, mucous and salivary amylase. The function of saliva is to lubricate food as it is chewed and swallowed and salivary amylase begins the digestion of starch.

2. Gastric juice secreted into the stomach from glands in its walls. Gastric juice contains pepsin that breaks down protein and hydrochloric acid to produce the acidic conditions under which this enzyme works best. In baby animals rennin to digest milk is also produced in the stomach.
3. Bile produced by the liver. It is stored in the gall bladder and secreted into the duodenum via the bile duct. (Note that the horse, deer, parrot and rat have no gall bladder). Bile is not a digestive enzyme. Its function is to break up large globules of fat into smaller ones so the fat splitting enzymes can gain access the fat molecules.

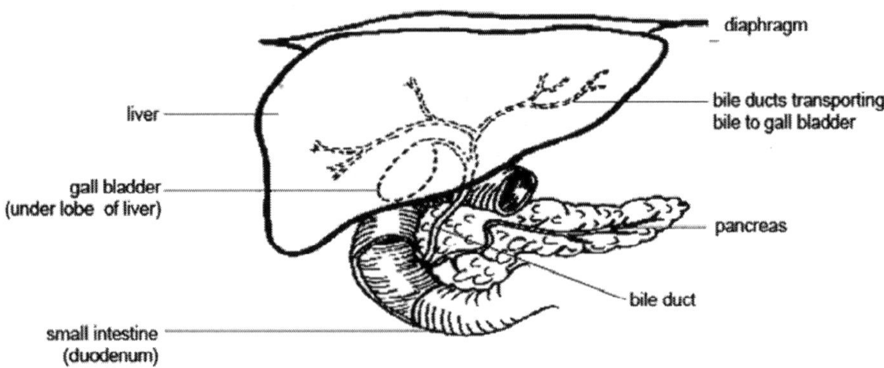

Figure 16: *The liver, gall bladder and pancreas*

Pancreatic Juice

The pancreas is a gland located near the beginning of the duodenum. In most animals it is large and easily seen but in rodents and rabbits it lies within the membrane linking the loops of the intestine (the mesentery) and is quite difficult to find. Pancreatic juice is produced in the pancreas. It flows into the duodenum and contains amylase for digesting starch, lipase for digesting fats and protease for digesting proteins.

Intestinal Juice

Intestinal juice is produced by glands in the lining of the small intestine. It contains enzymes for digesting disaccharides and proteins as well as mucus and salts to make the contents of the small intestine more alkaline so the enzymes can work.

Absorption

The small molecules produced by digestion are absorbed into the villi of the wall of the small intestine. The tiny finger-like projections of the villi increase the surface area for absorption. Glucose and amino acids pass directly through the wall into the blood stream by diffusion or active transport. Fatty acids and glycerol enter vessels of the lymphatic system (lacteals) that run up the centre of each villus.

The Liver

The liver is situated in the abdominal cavity adjacent to the diaphragm. It is the largest single organ of the body and has over 100 known functions. Its most important digestive functions are:

1. the production of bile to help the digestion of fats and
2. the control of blood sugar levels.

Glucose is absorbed into the capillaries of the villi of the intestine. The blood stream takes it directly to the liver via a blood vessel known as the hepatic portal vessel or vein.

The liver converts this glucose into glycogen which it stores. When glucose levels are low the liver can convert the glycogen back into glucose. It releases this back into the blood to keep the level of glucose constant. The hormone insulin, produced by special cells in the pancreas, controls this process.

Other functions of the liver include:

3. making vitamin A,
4. making the proteins that are found in the blood plasma (albumin, globulin and fibrinogen),
5. storing iron,
6. removing toxic substances like alcohol and poisons from the blood and converting them to safer substances,
7. producing heat to help maintain the temperature of the body.

Urinary System

Homeostasis

The cells of an animal can only remain healthy if the conditions are just right. The processes that take place in them are upset if the temperature is too high or too low, or if the fluid around or inside them is too acid or alkaline. Homeostasis is the name given to the processes that help keep the internal conditions constant even when external conditions change. The word means, "staying the same".

There are a number of organs in the body that play a part in maintaining homeostasis. For example, the skin helps keep the internal temperature of bird and mammals bodies within a narrow range even when the outside temperatures change; the lungs control the amount of carbon dioxide in the blood; the liver and pancreas work together to keep the amount of glucose in the blood within narrow limits, and the kidneys regulate the acidity and the concentration of water and salt in the blood. Hormones are chemicals that carry messages around the body in the blood and are central to many of the homeostatic processes mentioned above.

Water In The Body

Water is essential for living things to survive because all the chemical reactions within a body take place in a solution of water. An animal's body consists of up to 80% water. The exact proportion depends on the type of animal, its age, sex, health and whether or not it has had sufficient to drink. Generally animals do not survive a loss of more than 15% of their body water. In vertebrates almost 2/3rd of this water is in the cells (intracellular fluid). The rest is outside the cells (extracellular fluid) where it is found in the spaces around the cells (tissue fluid), as well as in the blood and lymph.

Maintaining Water Balance

Animals lose water through their skin and lungs, in the faeces and urine. These losses must be made up by water in food and drink and from the water that is a by-product of chemical reactions. If the animal does not manage to compensate for water loss the dissolved substances in the blood may become so concentrated they become lethal. To prevent this happening various mechanisms come into play as soon as the concentration of the blood increases. A part of the brain called the hypothalamus is in charge of these homeostatic processes. The most important is the feeling of thirst that is triggered by an increase in blood concentration. This stimulates an animal to find water and drink it.

The kidneys are also involved in maintaining water balance as various hormones instruct them to produce more concentrated urine and so retain some of the water that would otherwise be lost.

Desert Animals

Coping with water loss is a particular problem for animals that live in dry conditions. Some, like the camel, have developed great tolerance for dehydration. For example, under some conditions, camels can withstand the loss of one third of their body mass as water. They can also survive wide daily changes in temperature. This means they do not have to use large quantities of water in sweat to cool the body by evaporation.

Smaller animals are more able than large ones to avoid extremes of temperature or dry conditions by resting in sheltered more humid situations during the day and being active only at night.

The kangaroo rat is able to survive without access to any drinking water at all because it does not sweat and produces extremely concentrated urine. Water from its food and from chemical processes is sufficient to supply all its requirements.

Excretion

Animals need to excrete because they take in substances that are excess to the body's requirements and many of the chemical reactions in the body produce waste products. If these substances were not removed they would poison cells or slow down metabolism. All animals therefore have some means of getting rid of these wastes. The major waste products in mammals are carbon dioxide that is removed by the lungs, and urea that is produced when excess amino acids (from proteins) are broken down. Urea is filtered from the blood by the kidneys.

The Kidneys And Urinary System

The kidneys in mammals are bean-shaped organs that lie in the abdominal cavity attached to the dorsal wall on either side of the spine. An artery from the dorsal aorta called the renal artery supplies blood to them and the renal vein drains them.

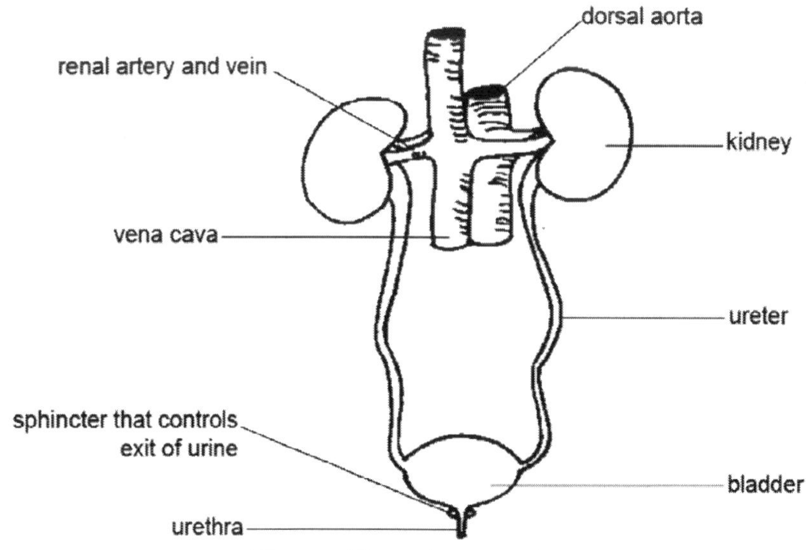

Figure: *The urinary system*

To the naked eye kidneys seem simple enough organs. They are covered by a fibrous coat or capsule and if cut in half lengthways (longitudinally) two distinct regions can be seen - an inner region or medulla and the outer cortex. A cavity within the kidney called the pelvis collects the urine and carries it to the ureter, which connects with the bladder where the urine is stored temporarily. Rings of muscle (sphincters) control the release of urine from the bladder and the urine leaves the body through the urethra.

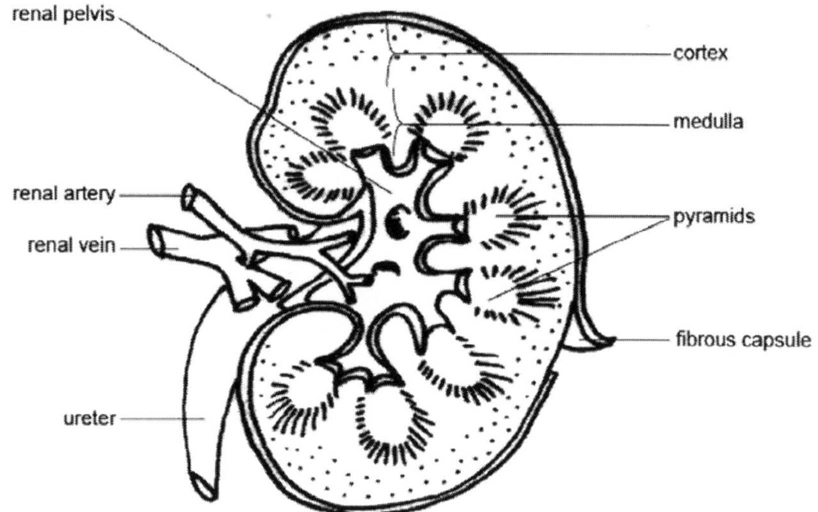

Figure: *The dissected kidney*

Kidney Tubules or Nephrons

It is only when you examine kidneys under the microscope that you find that their structure is not simple at all. The cortex and medulla are seen to be composed of masses of tiny tubes. These are called kidney tubules or nephrons. A human kidney consists of over a million of them.

At one end of each nephron, in the cortex of the kidney, is a cup shaped structure called the (Bowman's or renal) capsule. It surrounds a tuft of capillaries called the glomerulus that carries high-pressure blood. Together the glomerulus and capsule act as a blood-filtering device. The holes in the filter allow most of the contents of the blood through except the red and white cells and large protein molecules. The fluid flowing from the capsule into the rest of the kidney tubule is therefore very similar to blood plasma and contains many useful substances like water, glucose, salt and amino acids. It also contains waste products like urea.

Processes Occurring In The Nephron

After entering the glomerulus the filtered fluid flows along a coiled part of the tubule (the proximal convoluted tubule) to a looped portion (the Loop of Henle) and then to the collecting tube via a second length of coiled tube (the distal convoluted tubule). From the collecting ducts the urine flows into the renal pelvis and enters the ureter.

Note that the glomerulus, capsule and both coiled parts of the tubule are all situated in the cortex of the kidney while the loops of Henle and collecting ducts make up the medulla.

As the fluid flows along the proximal convoluted tubule useful substances like glucose, water, salts, potassium ions, calcium ions and amino acids are reabsorbed into the blood capillaries that form a network around the tubules. Many of these substances are transported by active transport and energy is required. In a separate process, some substances, particularly potassium, ammonium and hydrogen ions, and drugs like penicillin, are actively secreted into the distal convoluted tubule.

By the time the fluid has reached the collecting ducts these processes of absorption and secretion have changed the fluid originally filtered into the Bowman's capsule into urine. The main function of the collecting ducts is then to remove more water from the urine if necessary. These processes are summarised.

Normal urine consists of water, in which waste products such as urea and salts such as sodium chloride are dissolved. Pigments from the breakdown of red blood cells give urine its yellow colour.

The Production Of Concentrated Urine

Because of the high pressure of the blood in the glomerulus and the large size of the pores in the glomerulus/capsule-filtering device, an enormous volume of fluid passes into the kidney tubules. If this fluid were left as it is, the animal's body would be drained dry in 30 minutes. In fact, as the fluid flows down the tubule, over 90% of the water in it is reabsorbed. The main part of this reabsorption takes place in the collecting tubes.

The amount of water removed from the collecting ducts is controlled by a hormone called antidiuretic hormone (ADH) produced by the pituitary gland, situated at the base of the brain. When the blood becomes more concentrated, as happens when an animal is deprived of water, ADH is secreted and causes more water to be absorbed from the collecting ducts so that concentrated urine is produced. When the animal has drunk plenty of water and the blood is dilute, no ADH is secreted and no or little water is absorbed from the collecting ducts, so dilute urine is produced. In this way the concentration of the blood is controlled precisely.

Water Balance In Fish and Marine Animals

Fresh Water Fish: Although the skin of fish is more or less waterproof, the gills are very porous. The body fluids of fish that live in fresh water have a higher concentration of dissolved substances than the water in which they swim. In other words the body fluids of fresh water fish are hypertonic to the water. Water therefore flows into the body by osmosis. To stop the body fluids being constantly diluted fresh water fish produce large quantities of dilute urine.

Marine Fish

Marine fish like the sharks and dogfish have body fluids that have the same concentration of dissolved substances as the water (isotonic) have little problem with water balance. However, marine bony fish like red cod, snapper and sole, have body fluids with a lower concentration of dissolved substances than seawater (they are hypotonic to seawater). This means that water tends to flow out of their bodies by osmosis. To make up this fluid loss they drink seawater and get rid of the excess salt by excreting it from the gills.

Marine Birds

Marine birds that eat marine fish take in large quantities of salt and some only have access to seawater for drinking. Bird's kidneys are unable to produce very concentrated urine, so they have developed a salt gland. This excretes a concentrated salt solution into the nose to get rid of the excess salt.

Diabetes and The Kidney

There are two types of diabetes. The most common is called sugar diabetes or diabetes mellitus and is common in cats and dogs especially if they are overweight. It is caused by the pancreas secreting insufficient insulin, the hormone that controls the amount of glucose in the blood. If insulin secretion is inadequate, the concentration of glucose in the blood increases. Any increase in the glucose in the blood automatically leads to an increase in glucose in the fluid filtered into the kidney tubule. Normally the kidney removes all the glucose filtered into it, but these high concentrations swamp this removal mechanism and urine containing glucose is produced. The main symptoms of this type of diabetes are the production of large amounts of dilute urine containing glucose, and excessive thirst.

The second type of diabetes is called diabetes insipidus. The name comes from the main symptom, which is the production of large amounts of very dilute and "tasteless" urine. It occurs when the pituitary gland produces insufficient ADH, the hormone that stimulates water reabsorption from the kidney tubule. When this hormone is lacking, water is not absorbed and large amounts of dilute urine are produced. Because so much water is lost in the urine, animals with this form of diabetes can die if deprived of water for only a day or so.

Other Functions of the Kidney

The excretion of urea from the body and the maintenance of water balance, as described above, are the main functions of the kidney. However, the kidneys have other roles in keeping conditions in the body stable i.e. in maintaining homeostasis. These include:

- controlling the concentration of salt ions (Na+, K+, Cl-) in the blood by adjusting how much is excreted or retained;
- maintaining the correct acidity of the blood. Excess acid is constantly being produced by the normal chemical reactions in the body and the kidney eliminates this.

Normal Urine

Normal urine consists of water (95%), urea, salts (mostly sodium chloride) and pigments (mostly from bile) that give it its characteristic colour.

Abnormal Ingredients of Urine

If the body is not working properly, small amounts of substances not normally present may be found in the urine or substances normally present may appear in abnormal amounts.

- The presence of glucose may indicate diabetes.
- Urine with red blood cells in it is called haematuria, and may indicate inflammation of the kidney or urinary tract, cancer or a blow to the kidneys.
- Sometimes free haemoglobin is found in the urine. This indicates that the red blood cells in the blood have haemolysed (the membrane has broken down) and the haemoglobin has passed into the kidney tubules.
- The presence of white blood cells in the urine indicates there is an infection in the kidney or urinary tract.
- Protein molecules are usually too large to pass into the kidney tubule so no or only small amounts of proteins like albumin is normally found in urine. Large quantities of albumin indicate that the kidney tubules have been injured or the kidney has become diseased. High blood pressure also pushes proteins from the blood into the tubules.
- Casts are tiny cylinders of material that have been shed from the lining of the tubules and flushed out into the urine.
- Mucus is not usually found in the urine of healthy animals but is a normal constituent of horses' urine, giving it a characteristic cloudy appearance.

Tests can be carried out to identify any abnormal ingredients of urine. These tests are normally done by "stix", which are small plastic strips with absorbent ends impregnated with various chemicals. A colour change occurs in the presence of an abnormal ingredient.

Excretion in Birds

Birds' high body temperature and level of activity means that they need to conserve water. Birds therefore do not have a bladder and instead of excreting urea, which needs to be dissolved in large amounts of water, birds produce uric acid that can be discharged as a thick paste along with the feces. This is the white chalky part of the bird droppings that land on you or your car.

Nervous System

Coordination

Animals must be able to sense and respond to the environment in which they live if they are to survive. They need to be able to sense the temperature of their surroundings, for example, so they can avoid the hot sun. They must also be able to identify food and escape predators.

The various systems and organs in the body must also be linked so they work together. For example, once a predator has identified suitable prey it has to catch it. This involves coordinating the contraction of the muscle so the predator can run, there must then be an increased blood supply to the muscles to provide them with oxygen and nutrients. At the same time the respiration rate must increase to supply the oxygen and remove the carbon dioxide produced as a result of this increased activity. Once the prey has been caught and eaten, the digestive system must be activated to digest it.

The adjustment of an animal's response to changes in the environment and the complex linking of the various processes in the body that this response involves are called co-ordination. Two systems are involved in co-ordination in animals. These are the nervous and endocrine systems. The first operates via electrical impulses along nerve fibres and the second by releasing special chemicals or hormones into the bloodstream from glands.

Functions of the Nervous System

The nervous system has three basic functions:

1. *Sensory Function:* to sense changes (known as stimuli) both outside and within the body. For example the eyes sense changes in light and the ear responds to sound waves. Inside the body,

tretch receptors in the stomach indicate when it is full and chemical receptors in the blood vessels monitor the acidity of the blood.

2. *Integrative Function:* processing the information received from the sense organs. The impulses from these organs are analysed and stored as memory. The many different impulses from different sources are sorted, synchronised and co-ordinated and the appropriate response initiated. The power to integrate, remember and apply experience gives higher animals much of their superiority.
3. *Motor function:* The third function is the response to the stimuli that causes muscles to contract or glands to secrete.

All nervous tissue is made up of nerve cells or neurons. These transmit high-speed signals called nerve impulses. Nerve impulses can be thought of as being similar to an electric current.

The Neuron

Neurons are cells that have been adapted to carry nerve impulses. A typical neuron has a cell body containing a nucleus, one or more branching filaments called dendrites which conduct nerve impulses towards the cell body and one long fibre, an axon, that carries the impulses away from it. Many axons have a sheath of fatty material called myelin surrounding them. This speeds up the rate at which the nerve impulses travel along the nerve.

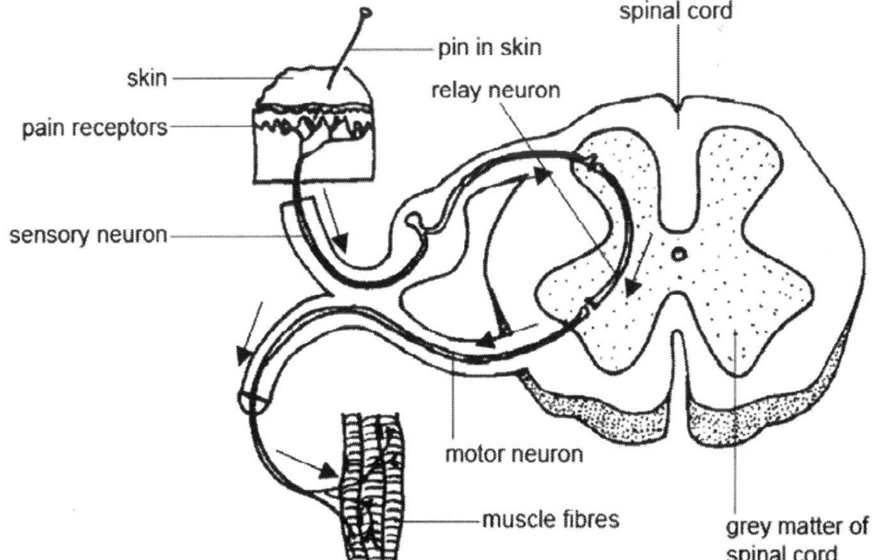

Figure: *The relationship between sensory, relay and motor neurons*

The cell body of neurons is usually located in the brain or spinal cord while the axon extends the whole distance to the organ that it supplies. The neuron carrying impulses from the spinal cord to the hind leg or tail of a horse, for example, can be several feet long. A nerve is a bundle of axons. A sensory neuron is a nerve cell that transmits impulses from a sense receptor such as those in the eye or ear to the brain or spinal cord. A motor neuron is a nerve cell that transmits impulses from the brain or spinal cord to a muscle or gland. A relay neuron connects sensory and motor neurons and is found in the brain or spinal cord.

Connections between Neurons

The connection between adjacent neurons is called a synapse. The two nerve cells do not actually touch here for there is a microscopic space between them. The electrical impulse in the neurone before the synapse stimulates the production of chemicals called neurotransmitters (such as acetylcholine), which are secreted into the gap.

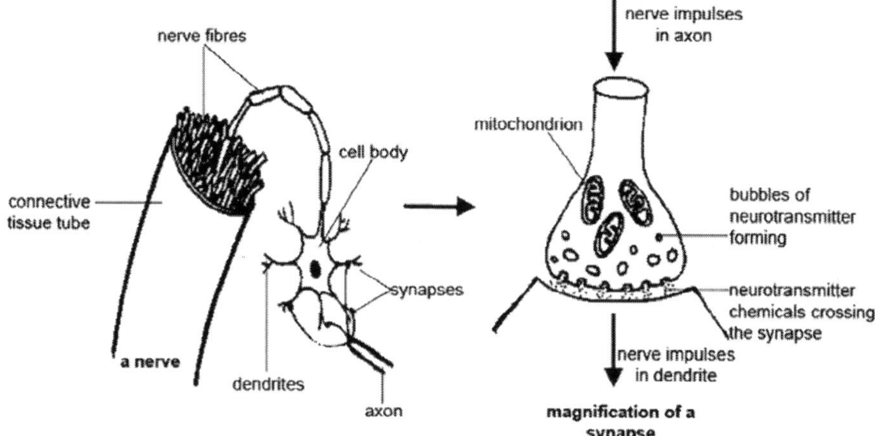

Figure: *A nerve and magnification of a synapse*

The neurotransmitter chemicals diffuse across the gap and when they contact the membrane of the next nerve cell they stimulate a new nervous impulse. After the impulse has passed the chemical is destroyed and the synapse is ready to receive the next nerve impulse.

Reflexes

A reflex is a rapid automatic response to a stimulus. When you accidentally touch a hot object and automatically jerk your hand away, this is a reflex action. It happens without you having to think about it. Animals automatically blink when an object approaches the eye and cats twist their bodies in the air when falling so they land on their paws. Swallowing, sneezing, and the constriction of the pupil of the eye in bright light are also all reflex actions.

The path taken by the nerve impulses in a reflex is called a reflex arc. Most reflex arcs involve only three neurons. The stimulus (a pin in the paw) stimulates the pain receptors of the skin, which initiate an impulse in a sensory neuron.

This travels to the spinal cord where it passes, by means of a synapse, to a connecting neuron called the relay neuron situated in the spinal cord. The relay neuron in turn makes a synapse with one or more motor neurons that transmit the impulse to the muscles of the limb causing them to contract and remove the paw from the sharp object. Reflexes do not require involvement of the brain although you are aware of what is happening and can, in some instances, prevent them happening. Animals are born with their reflexes. You can think of them as being wired in.

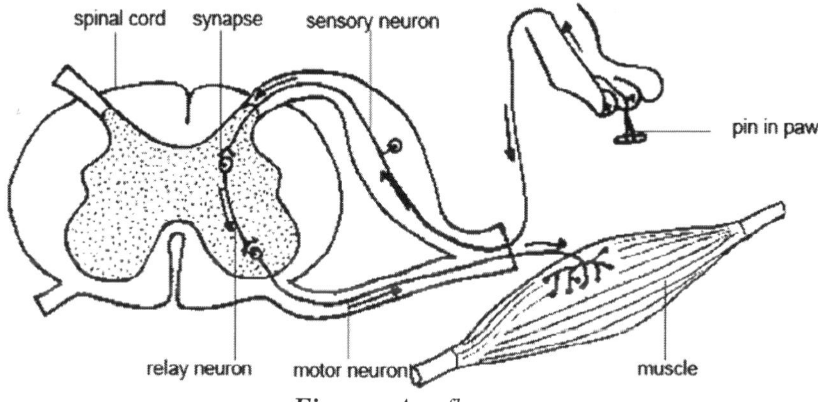

Figure: *A reflex arc*

Conditioned Reflexes

In most reflexes the stimulus and response are related. For example the presence of food in the mouth causes the salivary glands to release saliva. However, it is possible to train animals (and humans) to respond to different and often quite irrelevant stimuli. This is called a conditioned reflex.

A Russian biologist called Pavlov carried out the classic experiment to demonstrate such a reflex when he conditioned dogs to salivate at the sound of a bell ringing. Almost every pet owner can identify reflexes they have conditioned in their animals. Perhaps you have trained your cat to associate food with the opening of the fridge door or accustomed your dog to the routines you go through before taking them for a walk.

Parts of the Nervous System

When we describe the nervous system of vertebrates we usually divide it into two parts.

1. The central nervous system (CNS) which consists of the brain and spinal cord.
2. The peripheral nervous system (PNS) which consists of the nerves that connect to the brain and spinal cord (cranial and spinal nerves) as well as the autonomic (or involuntary) nervous system.

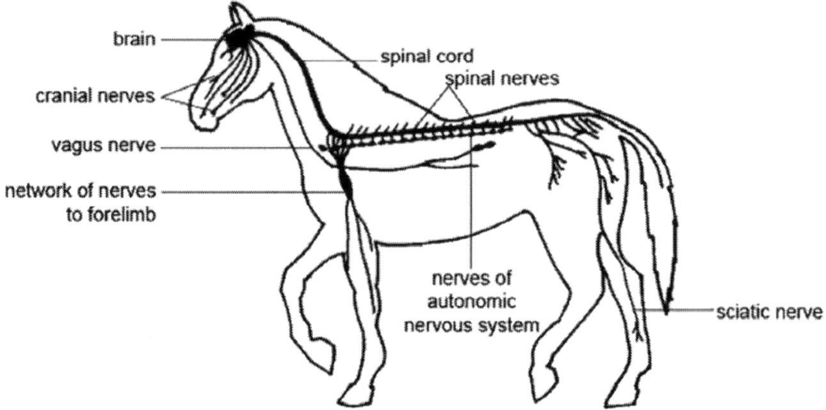

Figure: *The nervous system of a horse*

The Central Nervous System

The central nervous system consists of the brain and spinal cord. It acts as a kind of 'telephone exchange' where a vast number of cross connections are made. When you look at the brain or spinal cord some regions appear creamy white (white matter) and others appear grey (grey matter). White matter consists of masses of nerve axons and the grey matter consists of the nerve cells. In the brain the grey matter is on the outside and in the spinal cord it is on the inside.

The Brain

The major part of the brain lies protected within the sturdy "box" of skull called the cranium. Surrounding the fragile brain tissue (and spinal cord) are protective membranes called the meninges, and a crystal-clear fluid called cerebrospinal fluid, which protects and nourishes the brain tissue. This fluid also fills four cavities or ventricles that lie within the brain.

Brain tissue is extremely active and, even when an animal is resting, it uses up to 20% of the oxygen taken into the body by the lungs. The carotid artery, a branch off the dorsal aorta, supplies it with the oxygen and nutrients it requires. Brain damage occurs if brain tissue is deprived of oxygen for only 4-8 minutes.

The brain consists of three major regions:

1. the fore brain which includes the cerebral hemispheres, hypothalamus and pituitary gland;
2. the hind brain or brain stem, contains the medulla oblongata and pons and
3. the cerebellum or "little brain".

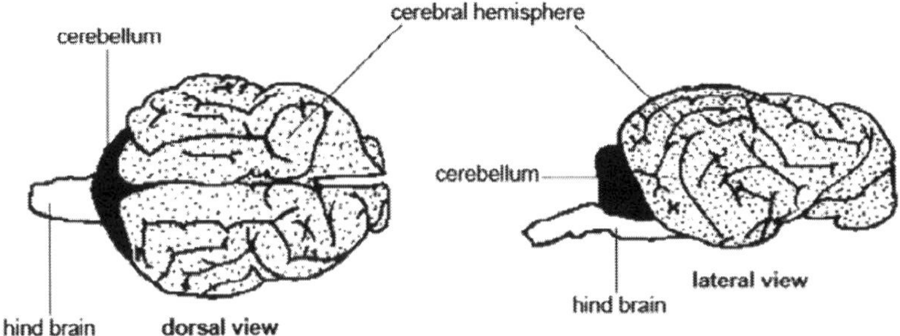

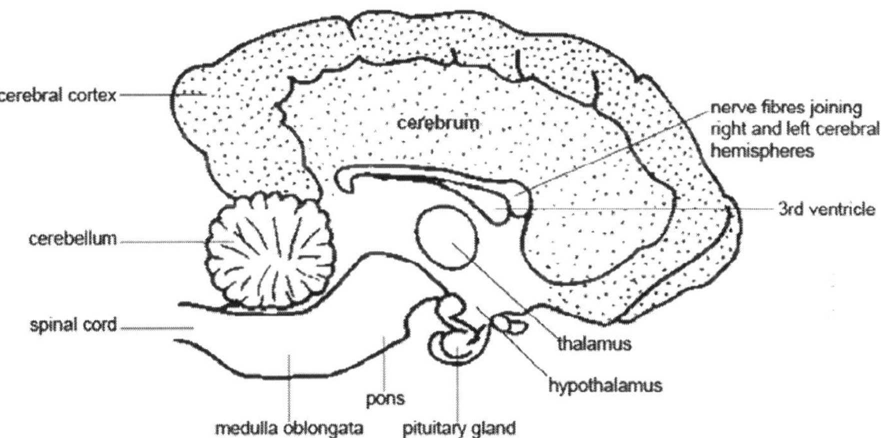

Figure: *Longitudinal section through the brain of a dog*

Mapping the Brain

In humans and some animals the functions of the different regions of the cerebral cortex have been mapped.

The Forebrain

The cerebral hemispheres are the masses of brain tissue that sit on the top of the brain. The surface is folded into ridges and furrows called sulci (singular sulcus). They make this part of the brain look rather like a very large walnut kernel. The two hemispheres are separated by a deep groove although they are connected internally by a thick bundle of nerve fibres. The outer layer of each hemisphere is called the cerebral cortex and this is where the main functions of the cerebral hemispheres are carried out.

The cerebral cortex is large and convoluted in mammals compared to other vertebrates and largest of all in humans because this is where the so-called “higher centres” concerned with memory, learning, reasoning and intelligence are situated.

Nerves from the eyes, ears, nose and skin bring sensory impulses to the cortex where they are interpreted. Appropriate voluntary movements are initiated here in the light of the memories of past events.

Different regions of the cortex are responsible for particular sensory and motor functions, e.g. vision, hearing, taste, smell, or moving the fore-limbs, hind-limbs or tail. For example, when a dog sniffs a scent, sensory impulses from the organ of smell in the nose pass via the olfactory (smelling) nerve to the olfactory centres of the cerebral hemispheres where the impulses are interpreted and co-ordinated.

In humans and some animals the functions of the different regions of the cerebral cortex have been mapped.

The hypothalamus is situated at the base of the brain and is connected by a “stalk” to the pituitary gland, the “master” hormone-producing gland. The hypothalamus can be thought of as the bridge between the nervous and endocrine (hormone producing) systems. It produces some of the hormones that are released from the pituitary gland and controls the release of others from it.

It is also an important centre for controlling the internal environment of the animal and therefore maintaining homeostasis. For example, it helps regulate the movement of food through the gut and the temperature, blood pressure and concentration of the blood. It is also responsible for the feeling of being hungry or thirsty and it controls sleep patterns and sex drive.

The Hindbrain

The medulla oblongata is at the base of the brain and is a continuation of the spinal cord. It carries all signals between the spinal cord and the brain and contains centres that control vital body functions like the basic rhythm of breathing, the rate of the heartbeat and the activities of the gut. The medulla oblongata also co-ordinates swallowing, vomiting, coughing and sneezing.

The Cerebellum

The cerebellum (little brain) looks rather like a smaller version of the cerebral hemispheres attached to the back of the brain. It receives impulses from the organ of balance (vestibular organ) in the inner ear and from stretch receptors in the muscles and tendons. By co-ordinating these it regulates muscle contraction during walking and running and helps maintain the posture and balance of the animal. When the cerebellum malfunctions it causes a tremor and uncoordinated movement.

The Spinal Cord

The spinal cord is a cable of nerve tissue that passes down the channel in the vertebrae from the hindbrain to the end of the tail. It becomes progressively smaller as paired spinal nerves pass out of the cord to parts of the body. Protective membranes or meninges cover the cord and these enclose cerebral spinal fluid.

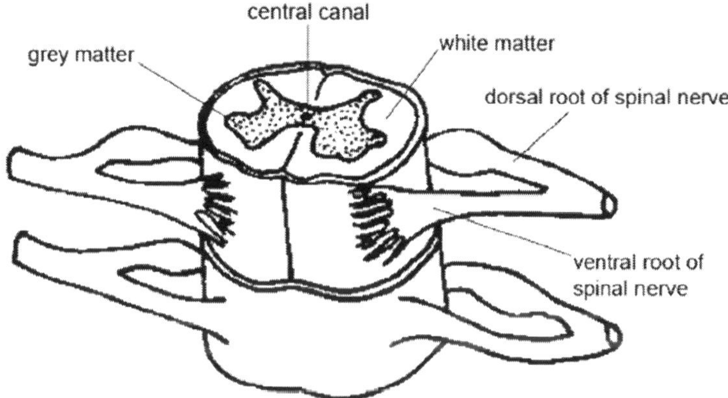

Figure: *The spinal cord*

If you cut across the spinal cord you can see that it consists of white matter on the outside and grey matter in the shape of an H or butterfly on the inside.

The Peripheral Nervous System

The peripheral nervous system consists of nerves that are connected to the brain (cranial nerves), and nerves that are connected to the spinal cord (spinal nerves). The autonomic nervous system is also part of the peripheral nervous system.

Cranial Nerves

There are twelve pairs of cranial nerves that come from the brain. Each passes through a hole in the cranium (brain case). The most important of these are the olfactory, optic, acoustic and vagus nerves.

The olfactory nerves - (smell) carry impulses from the olfactory organ of the nose to the brain.

The optic nerves - (sight) carry impulses from the retina of the eye to the brain. The auditory (acoustic) nerves - (hearing) carry impulses from the cochlear of the inner ear to the brain. The vagus nerve - controls the muscles that bring about swallowing. It also controls the muscles of the heart, airways, lungs, stomach and intestines.

Spinal Nerves

Spinal nerves connect the spinal cord to sense organs, muscles and glands in the body. Pairs of spinal nerves leave the spinal cord and emerge between each pair of adjacent vertebrae. The sciatic nerve is the largest spinal nerve in the body. It leaves the spinal cord as several nerves that join to form a flat band of nervous tissue. It passes down the thigh towards the hind leg where it gives off branches to the various muscles of this limb.

The Autonomic Nervous System

The autonomic nervous system controls internal body functions that are not under conscious control. For example when a prey animal is chased by a predator the autonomic nervous system automatically increases the rate of breathing and the heartbeat. It dilates the blood vessels that carry blood to the muscles, releases glucose from the liver, and makes other adjustments to provide for the sudden increase in activity. When the animal has escaped and is safe once again the nervous system slows down all these processes and resumes all the normal body activities like the digestion of food.

The nerves of the autonomic nervous system originate in the spinal cord and pass out between the vertebrae to serve the various organs. There are two main parts to the autonomic nervous system — the sympathetic system and the parasympathetic system. The sympathetic system stimulates the "flight, fright, fight" response that allows an animal to face up to an attacker or make a rapid departure. It increases the heart and respiratory rates, as well as the amount of blood flowing to the skeletal muscles while blood flow to less critical regions like the gut and skin is reduced. It also causes the pupils of the eyes to dilate. Note that the effects of the sympathetic system are similar to the effects of the hormone adrenaline. The parasympathetic system does the opposite to the sympathetic system. It maintains the normal functions of the relaxed body. These are sometimes known as the "housekeeping" functions. It promotes effective digestion, stimulates defaecation and urination and maintains a regular heartbeat and rate of breathing.

The Senses

The Sense Organs

Sense organs allow animals to sense changes in the environment around them and in their bodies so that they can respond appropriately.

They enable animals to avoid hostile environments, sense the presence of predators and find food. Animals can sense a wide range of stimuli that includes, touch, pressure, pain, temperature, chemicals, light, sound, movement and position of the body. Some animals can sense electric and magnetic fields. All sense organs respond to stimuli by producing nerve impulses that travel to the brain via a sensory nerve. The impulses are then processed and interpreted in the brain as pain, sight, sound, taste etc.

The senses are often divided into two groups:

1. The general senses of touch, pressure, pain and temperature that are distributed fairly evenly through the skin. Some are found in muscles and within joints.
2. The special senses which include the senses of smell, taste, sight, hearing and balance. The special sense organs may be quite complex in structure.

Touch and Pressure

Within the dermis of the skin are numerous modified nerve endings that are sensitive to touch and pressure. The roots of hairs may also be well supplied with sensory receptors that inform the animal that it is in contact with an object. Whiskers are specially modified hairs.

Pain

Receptors that sense pain are found in almost every tissue of the body. They tell the animal that tissues are dangerously hot, cold, compressed or stretched or that there is not enough blood flowing in them. The animal may then be able to respond and protect itself from further damage

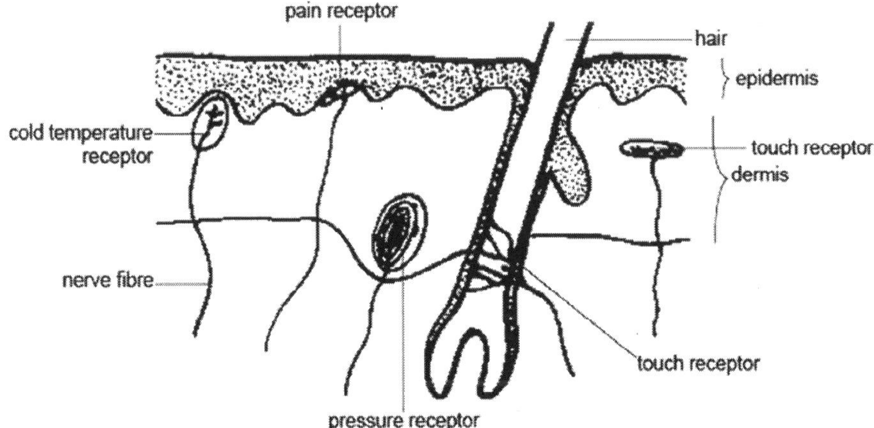

Figure: *The general senses in the skin*

Temperature

Nerve endings in the skin respond to hot and cold stimuli.

Awareness of Limb Position

There are sense organs in the muscles, tendons and joints that send continuous impulses to the brain that tell it where each limb is. This information allows the animal to place its limbs accurately and know their exact position without having to watch them.

Smell

Animals use the sense of smell to locate food, mark territory, identify their own offspring and the presence and sexual condition of a potential mate. The organ of smell (olfactory organ) is located in the nose and responds to chemicals in the air. It consists of modified nerve cells that have several tiny hairs on the surface. These emerge from the epithelium on the roof of the nose cavity into the mucus that lines it. As the animal breathes, chemicals in the air dissolve in the mucus. When the sense cell responds to a particular molecule, it fires an impulse that travels along the olfactory nerve to the brain where it is interpreted as an odour.

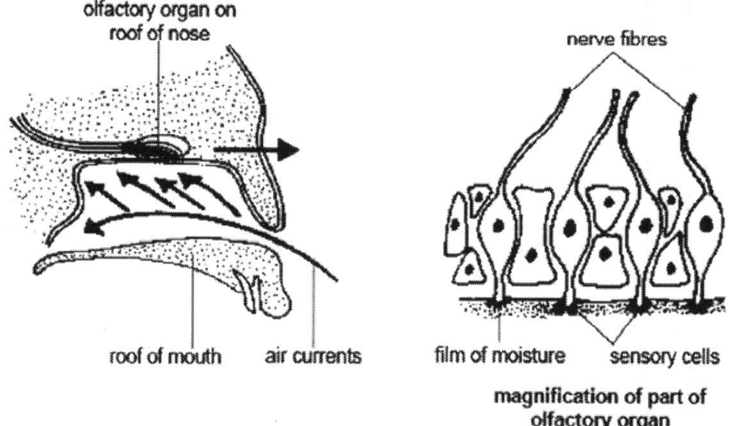

Figure: *The olfactory organ - the sense of smell*

The olfactory sense in humans is rudimentary compared to that of many animals. Carnivores that hunt have a very highly developed sensitivity to scents. For example a polar bear can smell out a dead seal 20 km away and a bloodhound can distinguish between the trails of different people although it may sometimes be confused by the criss-crossing trail of identical twins.

Snakes and lizards detect odours by means of Jacobson's organ. This is situated on the roof of the mouth and consists of pits containing sensory cells. When snakes flick out their forked tongues they are smelling the air by carrying the molecules in it to the Jacobson's organ.

Taste

The sense of taste allows animals to detect and identify dissolved chemicals. In reptiles, birds, and mammals the taste receptors (taste buds) are found mainly to the upper surface of the tongue. They consist of pits containing sensory cells arranged rather like the segments of an orange. Each receptor cell has a tiny "hair" that projects into the saliva to sense the chemicals dissolved in it. The sense of taste is quite restricted. Humans can only distinguish four different tastes (sweet, sour, bitter and salt) and what we normally think of as taste is mainly the sense of smell. Food is quite tasteless when the nose is blocked and cats often refuse to eat when this happens.

Sight

The eyes are the organs of sight. They consist of spherical eyeballs situated in deep depressions in the skull called the orbits. They are attached to the wall of the orbit by six muscles, which move the eyeball. Upper and lower eyelids cover the eyes during sleep and protect them from foreign objects or too much light, and spread the tears over their surface.

The nictitating membrane or hawis a transparent sheet that moves sideways across the eye from the inner corner, cleansing and moistening the cornea without shutting out the light. It is found in birds, crocodiles, frogs and fish as well as marsupials like the kangaroo. It is rare in mammals but can be seen in cats and dogs by gently opening the eye when it is asleep. Eyelashes also protect the eyes from the sun and foreign objects.

Structure of the Eye

Lining the eyelids and covering the front of the eyeball is a thin epithelium called the conjunctiva. Conjunctivitis is inflammation of this membrane. Tear glands that open just under the top eyelid secrete a salty solution that keeps the exposed part of the eye moist, washes away dust and contains an enzyme that destroys bacteria.

The wall of the eyeball is composed of three layers. From the outside these are the sclera, the choroid and the retina.

The sclera is a tough fibrous layer that protects the eyeball and gives it rigidity. At the front of the eye the sclera is visible as the "white" of the eye, which is modified as the transparent cornea through which the light rays have to pass to enter the eye. The cornea helps focus light that enters the eye.

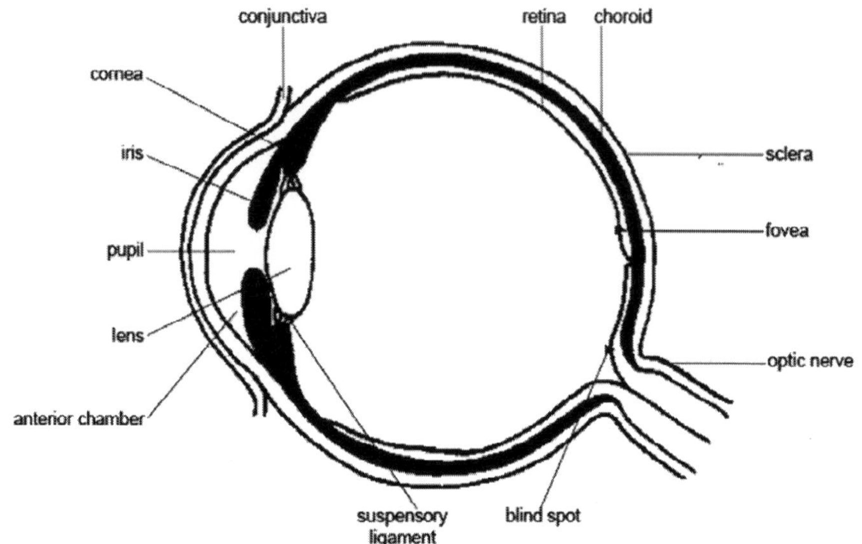

Figure: *The structure of the eye*

The choroid lies beneath the sclera. It contains a network of blood vessels that supply the eye with oxygen and nutrients. Its inner surface is highly pigmented and absorbs stray light rays. In nocturnal animals like the cat and possum this highly pigmented layer reflects light as a means of conserving light.

This is what makes them shine when caught in car headlights. At the front of the eye the choroid becomes the iris. This is the coloured part of the eye that controls the amount of light entering the pupil of the eye. In dim light the pupil is wide open so as much light as possible enters while in bright light the pupils contract to protect the retina from damage by excess light.

The pupil in most animals is circular but in many nocturnal animals it is a slit that can close completely. This helps protect the extra-sensitive light sensing tissues of animals like the cat and possum from bright sunlight.

The inner layer lining the inside of the eye is the retina. This contains the light sensing cells called rods and cones.

The rod cells are long and fat and are sensitive to dim light but cannot detect colour. They contain large amounts of a pigment that changes when exposed to light. This pigment comes from vitamin A found in carrots etc. A deficiency of this vitamin causes night blindness. So your mother was right when she told you to eat your carrots as they would help you see in the dark.

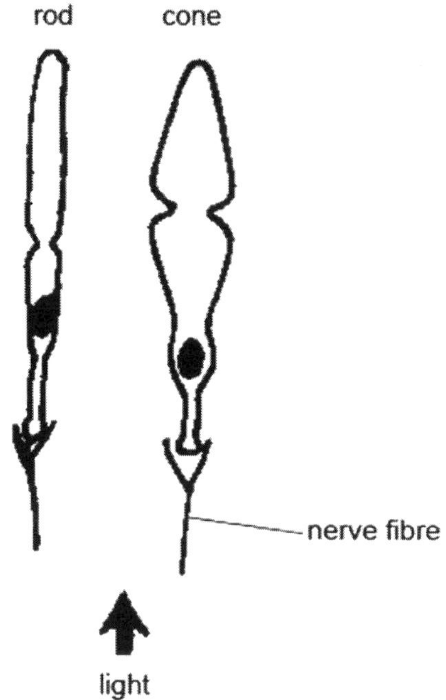

Figure: *A rod and cone from the retina*

The cone cells provide colour vision and allow animals to see details. Most are found in the centre of the retina and they are most densely concentrated in a small area called the fovea. This is the area of sharpest vision, where the words you are reading at this moment are focussed on your retina.

The nerve fibres from the cells of the retina join and leave the eye via the optic nerve. There are no rods or cones here and it is a blind spot. The optic nerve passes through the back of the orbit and enters the brain.

The lens is situated just behind the pupil and the iris. It is a crystalline structure with no blood vessels and is held in position by a ligament.

This is attached to a muscle, which changes the shape of the lens so both near and distant objects can be focussed by the eye. This ability to change the focus of the lens is called accommodation. In many mammals the muscles that bring about accommodation are poorly developed, Rats, cows and dogs, for example, are thought to be unable to focus clearly on near objects.

In old age and certain diseases the lens may become cloudy resulting in blurred vision. This is called a cataract. Within the eyeball are two cavities, the anterior and posterior chambers, separated by the lens. They contain fluids the aqueous and vitreous humours respectively, that maintain the shape of the eyeball and help press the retina firmly against the choroid so clear images are seen.

How the Eye Sees

Eyes work quite like a camera. Light rays from an object enter the eye and are focused on the retina (the "film") at the back of the eye. The cornea, the lens and the fluid within the eye all help to focus the light. They do this by bending the light rays so that light from the object falls on the retina.

This bending of light is called refraction. The light stimulates the light sensitive cells of the retina and nerve impulses are produced that pass down the optic nerve to the brain.

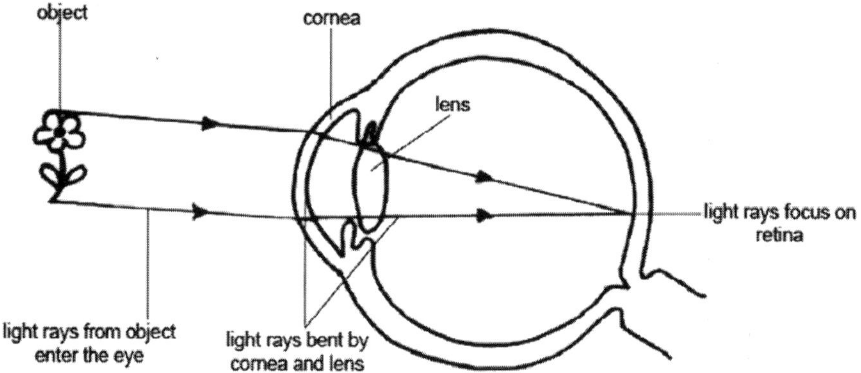

Figure: *How the light travels from the object to the retina of the eye*

Colour Vision in Animals

As mentioned before, the retina has two different kinds of cells that are stimulated by light - '*rods*' and '*cones*'. In humans and higher primates like baboons and gorillas the rods function in dim light and do not perceive colour, while the cones are stimulated by bright light and perceive details and colour.

Other mammals have very few cones in their retinas and it is believed that they see no or only a limited range of colour. It is, of course, difficult to find out exactly what animals do see. It is thought that deer, rats, and rabbits and nocturnal animals like the cat are colour-blind, and dogs probably see green and blue. Some fish and most birds seem to have better colour vision than humans and they use colour, often very vivid ones, for recognising each other as well as for courtship and protection.

Binocular Vision

Animals like cats that hunt have eyes placed on the front of the head in such a way that both eyes see the same wide area but from slightly different angles. This is called binocular vision. Its main advantage is that it enables the animals to estimate the distance to the prey so they can chase it and pounce accurately.

In contrast plant-eating prey animals like the rabbit and deer need to have a wide panoramic view so they can see predators approaching. They therefore have eyes placed on the side of the head, each with its own field of vision. They have only a very small area of binocular vision in front of the head but are extremely sensitive to movement.

Hearing

Animals use the sense of hearing for many different purposes. It is used to sense danger and enemies, to detect prey, to identify prospective mates and to communicate within social groups. Some animals (e.g. most bats and dolphins) use sound to "see" by echolocation. By sending out a cry and interpreting the echo, they sense obstacles or potential prey.

Structure of the Ear

Most of the ear, the organ of hearing, is hidden from view within the boney skull. It consists of three main regions: the outer ear, the middle ear and the inner ear.

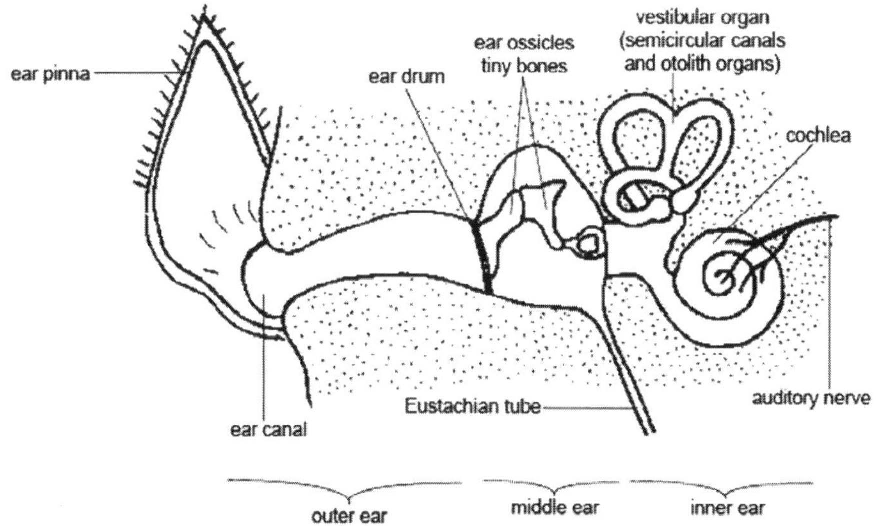

Figure: *The ear*

The outer ear consists of an ear canal leading inwards to a thin membrane known as the eardrumour tympanic membrane that stretches across the canal. Many animals have an external ear flap or pinna to collect and funnel the sound into the ear canal. The pinnae (plural of pinna) usually face forwards on the head but many animals can swivel them towards the source of the sound.

In dogs the ear canal is long and bent and often traps wax or provides an ideal habitat for mites, yeast and bacteria.

The middle ear consists of a cavity in the skull that is connected to the pharynx (throat) by a long narrow tube called the Eustachian tube. This links the middle ear to the outside air so that the air pressure on both sides of the eardrum can be kept the same. Everyone knows the uncomfortable feeling (and affected hearing) that occurs when you drive down a steep hill and the unequal air pressures on the two sides of the eardrum cause it to distort. The discomfort is relieved when you swallow because the Eustachian tubes open and the pressure on either side equalises.

Within the cavity of the middle ear are three of the smallest bones in the body, the auditory ossicles. They are known as the hammer, the anvil and the stirrup because of their resemblance to the shape of these objects. These tiny bones articulate (move against) each other and transfer the vibrations of the eardrum to the membrane covering the opening to the inner ear.

The inner ear is a complicated series of fluid-filled tubes imbedded in the bone of the skull. It consists of two main parts. These are the cochlea where sound waves are converted to nerve impulses and the vestibular organ that is associated with the sense of balance and has no role in hearing.

The cochlea looks rather like a coiled up snail shell. Within it there are specialised cells with fine hairs on their surface that respond to the movement of the fluid within the cochlea by producing nervous impulses that travel to the brain along the auditory nerve.

How The Ear Hears

Sound waves can be thought of as vibrations in the air. They are collected by the ear pinna and pass down the ear canal where they cause the eardrum to vibrate. (An interesting fact is that when you are listening to someone speaking your eardrum vibrates at exactly the same rate as the vocal cords of the person speaking to you).

The vibration of the eardrum sets the three tiny bones in the middle ear moving against each other so that the vibration is transferred to the membrane covering the opening to the inner ear. As well as transferring the vibration, the tiny ear bones also amplify it. The three tiny bones are called the stirrup, anvil, and hammer. They were called such of their form. In the human ear this amplification is about 20 times while in desert-dwelling animals like the kangaroo rat it is 100 times. This acute hearing warns them of the approach of predators like owls and snakes, even in the dark.

The vibration causes waves in the fluid in the inner ear that pass down the cochlea. These waves stimulate the tiny hair cells to produce nerve impulses that travel via the auditory nerve to the cerebral cortex of the brain where they are interpreted as sound.

To summarise: The route sound waves take as they pass through the ear is: External ear | tympanic membrane | ear ossicles | inner ear |cochlear | hair cells

The hair cells generate a nerve impulse that travels down the auditory nerve to the brain.

Remember that sound waves do not pass along the Eustachian tube. Its function is to equalise the air pressure on either side of the tympanic membrane.

Balance

The vestibular organ of the inner ear helps an animal maintain its posture and keep balanced by monitoring the movement and position of the head. It consists of two structures - the semicircular canals and the otolith organs.

The semicircular canals respond to movement of the body. They tell an animal whether it is moving up or down, left or right. They consist of three canals set in three different planes at right angles to each other so that movement in any direction can be registered. The canals contain fluid and sense cells with fine hairs that project into the fluid. When the head moves the fluid swirls in the canals and stimulates the hair cells. These send nerve impulses along the vestibular nerve to the cerebellum.

Note that the semicircular canals register acceleration and deceleration as well as changes in direction but do not respond to movement that is at a constant speed.

The otolith organs are sometimes known as gravity receptors. They tell you if your head is tilted or if you are standing on your head. They consist of bulges at the base of the semi circular canals that contain hair cells that are covered by a mass of jelly containing tiny pieces of chalk called otoliths. When the head is tilted, or moved suddenly, the otoliths pull on the hair cells, which produce a nerve impulse. This travels down the vestibular nerve to the cerebellum. By coordinating the nerve impulses from the semicircular canals and otolith organs the cerebellum helps the animal keep its balance.

- Receptors for touch, pressure, pain and temperature are found in the skin. Receptors in the muscles, tendons and joints inform the brain of limb position.
- The olfactory organ in the nose responds to chemicals in the air i.e. smell.
- Taste buds on the tongue respond to a limited range of chemicals dissolved in saliva.
- The eyes are the organs of sight. Spherical eyeballs situated in orbits in the skull have walls composed of 3 layers.
- The tough outer sclera protects and holds the shape of the eyeball. At the front it becomes visible as the white of the eye and the transparent cornea that allows light to enter the eye.
- The middle layer is the choroid. It most animals it absorbs stray light rays but in nocturnal animal it is reflective to conserve light. At the front of the eye it becomes the iris with muscles to control the size of the pupil and hence the amount of light entering the eye.
- The inner layer is the retina containing the light receptor cells: the rods for black and white vision in dim light and the cones for colour and detailed vision. Nerve impulses generated by these cells leave the eye for the brain via the optic nerve.
- The lens (with the cornea) helps focus the light rays on the retina. Muscles alter the shape of the lens to allow near and far objects to be focussed.
- Aqueous humour fills the space immediately behind the cornea and keeps it in shape and vitreous humour, a transparent jelly-like substance, fills the space behind the lens allowing light rays to pass through to the retina.
- The ear is the organ of hearing and balance.
- The external pinna helps funnel sound waves into the ear and locate the direction of the sound. The sound waves travel down the external ear canal to the eardrum or tympanic membrane

causing it to vibrate. This vibration is transferred to the auditory ossicles of the middle ear which themselves transfer it to the inner ear. Here receptors in the cochlea respond by generating nerve impulses that travel to the brain via the auditory (acoustic) nerve.

- The Eustachian tube connects the middle ear with the pharynx to equalise air pressure on either side of the tympanic membrane.
- The vestibular organ of the inner ear is concerned with maintaining balance and posture. It consists of the semicircular canals and the otolith organs.

Chapter 6

Anatomy and Physiology of Animals Main Organs

The Cell

The cell is the basic building block of living organisms. Bacteria and the parasite that causes malaria consist of single cells, while plants and animals are made up of trillions of cells. Most cells are spherical or cube shaped but some are a range of different shapes.

Most cells are so small that a microscope is needed to see them, although a few cells, e.g. the ostrich's egg, are so large that they could make a meal for several people.

A normal cell is about 0.02 of a millimetre (0.02mm) in diametre. (Small distances like this are normally expressed in micrometres or microns (ìm). Note there are 1000 ìms in every mm).

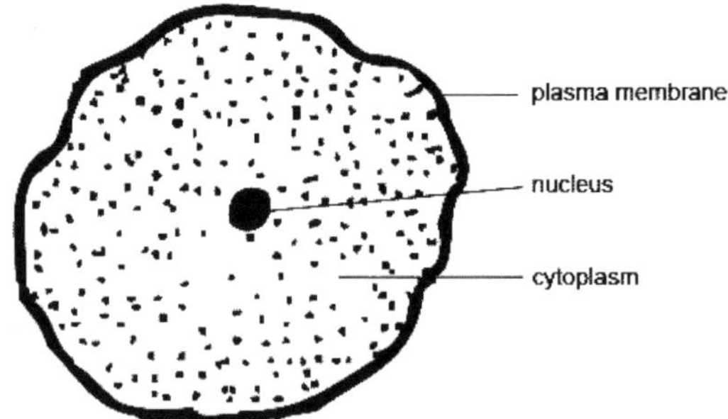

Figure: *An animal cell*

When you look at a typical animal cell with a light microscope it seems quite simple with only a few structures visible. Three main parts can be seen:

- an outer cell wall or plasma membrane,
- an inner region called the cytoplasm and
- the nucleus

However, when you use an electron microscope to increase the magnification many thousands of times you see that these seemingly simple structures are incredibly complex, each with its own specialised function. For example the plasma membrane is seen to be a double layer and the cytoplasm contains many special structures called organelles (meaning little organs) which are described below. A drawing of the cell as seen with an electron microscope.

The Plasma Membrane

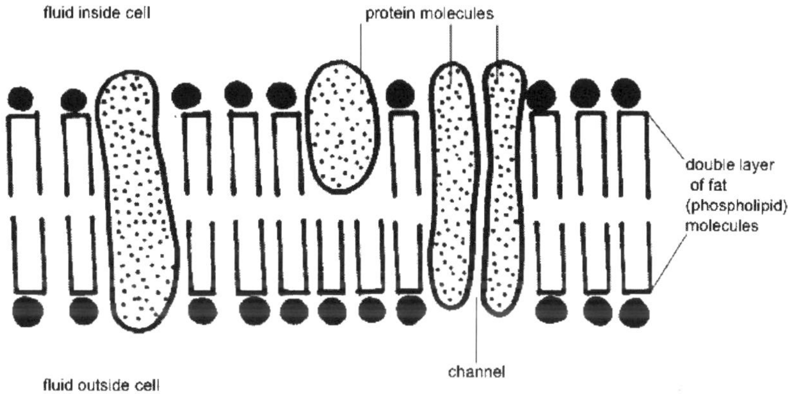

Figure: *The structure of the plasma membrane*

The thin plasma membrane surrounds the cell, separating its contents from the surroundings and controlling what enters and leaves the cell. The plasma membrane is composed of two main molecules, fats (in fact phospholipids) and proteins. The fats are arranged in a double layer with the large protein molecules dotted about in the membrane. Some of the protein molecules form tiny channels in the membrane while others help transport substances from one side of the membrane to the other.

How Substances Move Across the Plasma Membrane

Substances need to pass through the membrane to enter or leave the cell and they do so in a number of ways. Some of these processes require no energy i.e. they are passive, while others require energy i.e. they are active.

Passive processes include: a) diffusion and b) osmosis, while active processes include: c) active transport, d) phagocytosis, e) pinocytosis and f) exocytosis. These will be described below.

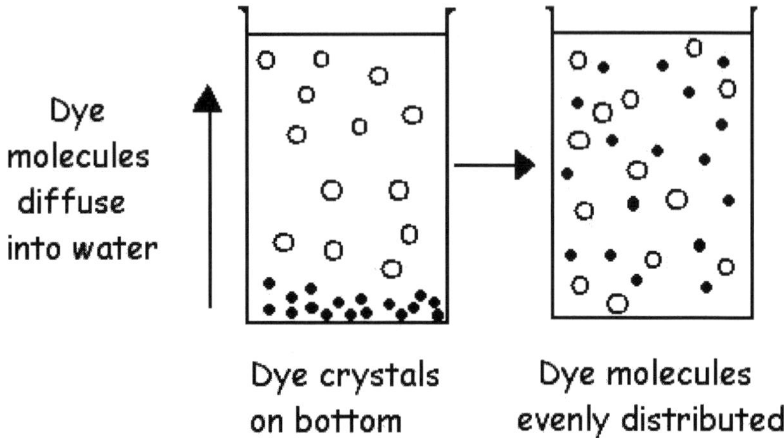

Figure: *Diffusion in a liquid*

Diffusion: Although you may not know it, you are already familiar with the process of diffusion. It is diffusion that causes a smell (expensive perfume or smelly socks) in one part of the room to gradually move through the room so it can be smelt on the other side. Diffusion occurs in the air and in liquids.

What happens when a few crystals of a dark purple dye called potassium permanganate are dropped into a beaker of water. The dye molecules diffuse into the water moving from high to low concentrations so they become evenly distributed throughout the beaker.

In the body, diffusion causes molecules that are in a high concentration on one side of the cell membrane to move across the membrane until they are present in equal concentrations on both sides. It takes place because all molecules have an inbuilt vibration that causes them to move and collide until they are evenly distributed. It is an absolutely natural process that requires no added energy.

Small molecules like oxygen, carbon dioxide, water and ammonia as well as fats, diffuse directly through the double fat layer of the membrane. The small molecules named above as well as a variety of charged particles (ions) also diffuse through the protein-lined channels. Larger molecules like glucose attach to a carrier molecule that aids their diffusion through the membrane. This is called facilitated diffusion.

In the animal's body diffusion is important for moving oxygen and carbon dioxide between the lungs and the blood, for moving digested food molecules from the gut into the blood and for the removal of waste products from the cell.

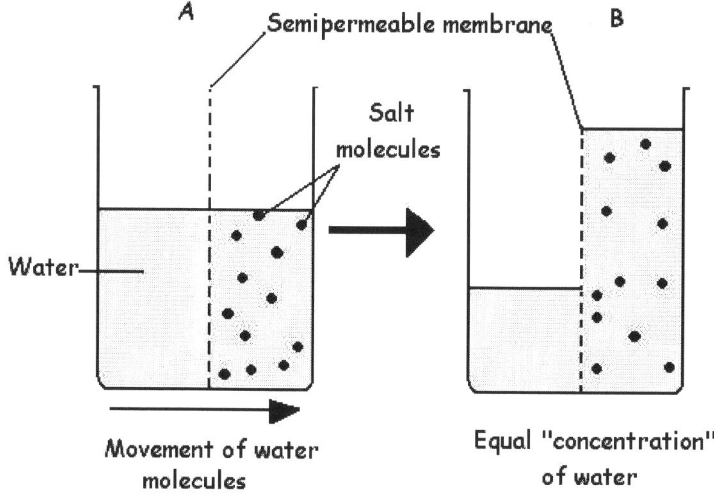

Figure: *Osmosis*

Osmosis: Although the word may be unfamiliar, you are almost certainly acquainted with the effects of osmosis. It is osmosis that plumps out dried fruit when you soak it before making a fruit cake or makes that wizened old carrot look almost like new when you soak it in water. Osmosis is in fact the diffusion of water across a membrane that allows water across but not larger molecules. This kind of membrane is called a semi-permeable membrane.

It shows a container divided into two parts by an artificial semi-permeable membrane. Water is poured into one part while a solution containing salt is poured into the other part. Water can cross the membrane but the salt cannot. The water crosses the semi-permeable membrane by diffusion until there is an equal amount of water on both sides of the membrane. The effect of this would be to make the salt solution more diluted and cause the level of the liquid in the right-hand side of the container to rise so it looked like side B. This movement of water across the semi-permeable membrane is called osmosis. It is a completely natural process that requires no outside energy.

Although it would be difficult to do in practice, imagine that you could now take a plunger and push down on the fluid in the right-hand side of container B so that it flowed back across the semi-permeable membrane until the level of fluid on both sides was equal again. If you could measure the pressure required to do this, this would be equal to the osmotic pressure of the salt solution. (This is a rather advanced concept at this stage but you will meet this term again when you study fluid balance later in the course).

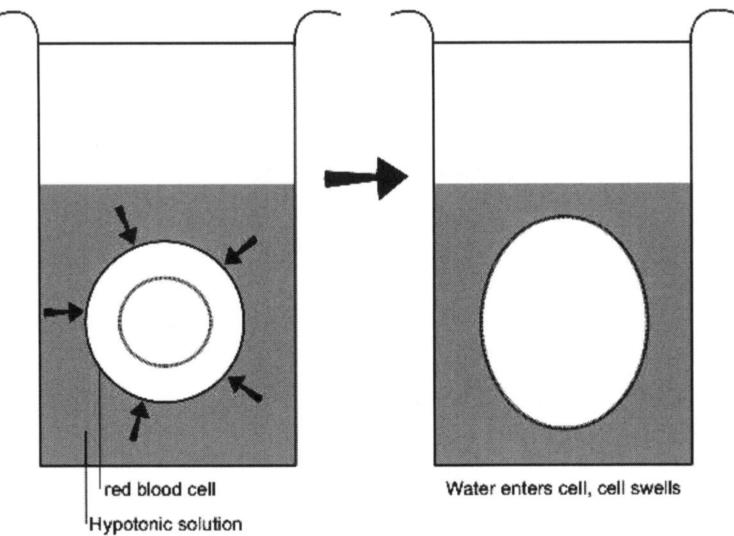

Figure: *Osmosis in red cells placed in a hypotonic solution*

The plasma membrane of cells acts as a semi-permeable membrane. If red blood cells, for example, are placed in water, the water crosses the membrane to make the amount of water on both sides of it equal. This means that the water moves into the cell causing it to swell. This can occur to such an extent that the cell actually bursts to release its contents. This bursting of red blood cells is called haemolysis. In a situation such as this when the solution on one side of a semi-permeable membrane has a lower concentration than that on the other side, the first solution is said to be hypotonic to the second.

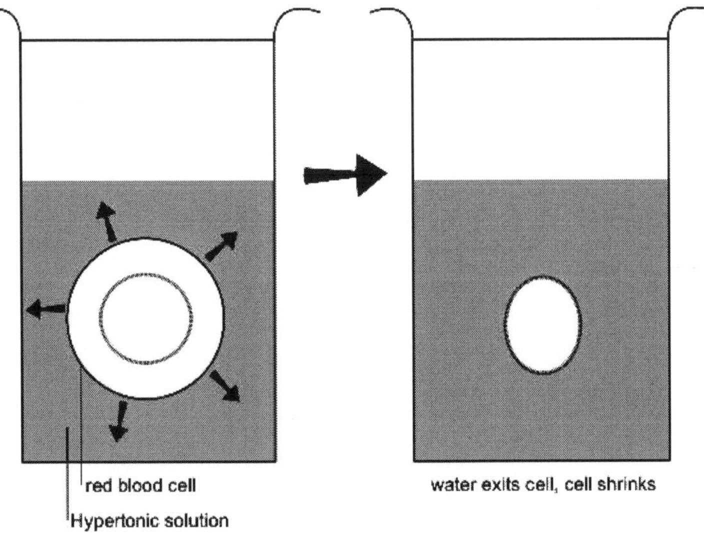

Figure: *Osmosis in red cells placed in a hypertonic solution*

Now think what would happen if red blood cells were placed in a salt solution that has a higher salt concentration than the solution within the cells. Such a bathing solution is called a hypertonic solution. In this situation the "concentration" of water within the cells would be higher than that outside the cells. Osmosis (diffusion of water) would then occur from the inside of the cells to the outside solution, causing the cells to shrink.

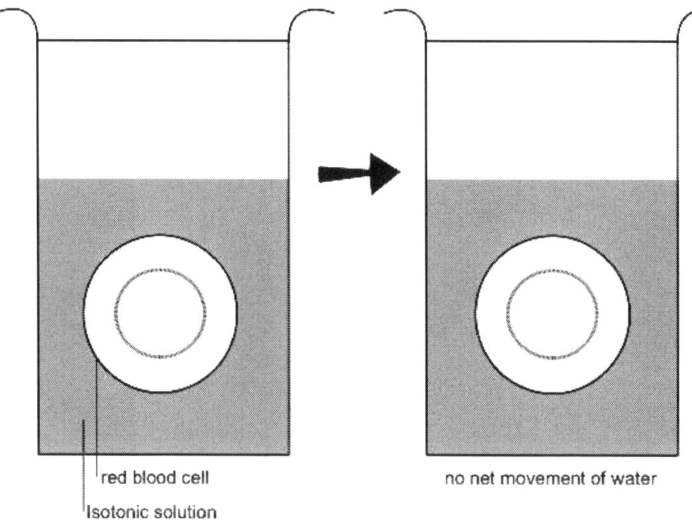

Figure: *Red cells placed in an isotonic solution*

A solution that contains 0.9% salt has the same concentration as body fluids and the solution within red cells. Cells placed in such a solution would neither swell nor shrink. This solution is called an isotonic solution. This strength of salt solution is often called normal saline and is used when replacing an animal's body fluids or when cells like red blood cells have to be suspended in fluid. Remember - osmosis is a special kind of diffusion. It is the diffusion of water molecules across a semi-permeable membrane. It is a completely passive process and requires no energy.

Sometimes it is difficult to remember which way the water molecules move. Although it is not strictly true in a biological sense, many students use the phrase "SALT SUCKS" to help them remember which way water moves across the membrane when there are two solutions of different salt concentrations on either side. As we have seen water moves in and out of the cell by osmosis. All water movement from the intestine into the blood system and between the blood capillaries and the fluid around the cells (tissue or extra cellular fluid) takes place by osmosis. Osmosis is also important in the production of concentrated urine by the kidney.

Active Transport: When a substance is transported from a low concentration to a high concentration i.e. uphill against the concentration gradient, energy has to be used. This is called active transport.

Active transport is important in maintaining different concentrations of the ions sodium and potassium on either side of the nerve cell membrane. It is also important for removing valuable molecules such as glucose, amino acids and sodium ions from the urine.

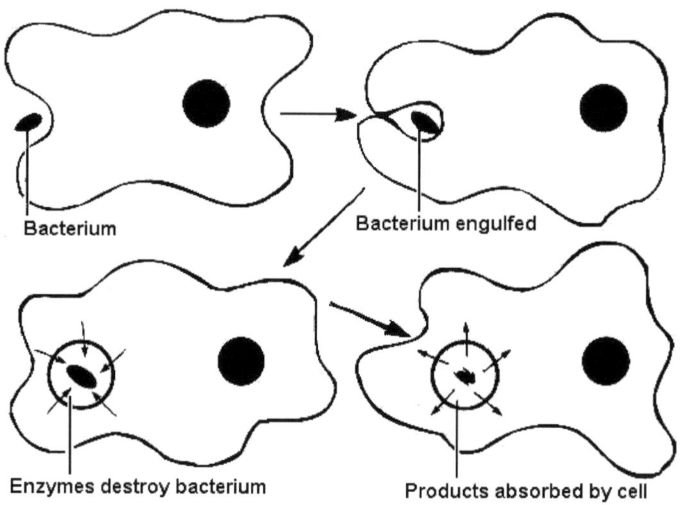

Figure: *Phagocytosis*

Phagocytosis: Phagocytosis is sometimes called "cell eating". It is a process that requires energy and is used by cells to move solid particles like bacteria across the plasma membrane. Finger-like projections from the plasma membrane surround the bacteria and engulf them as shown. Once within the cell, enzymes produced by the lysosomes of the cell (described later) destroy the bacteria.

The destruction of bacteria and other foreign substance by white blood cells by the process of phagocytosis is a vital part of the defence mechanisms of the body.

Pinocytosis: Pinocytosis or "cell drinking" is a very similar process to phagocytosis but is used by cells to move fluids across the plasma membrane. Most cells carry out pinocytosis.

Exocytosis: Exocytosis is the process by means of which substances formed in the cell are moved through the plasma membrane into the fluid outside the cell (or extra-cellular fluid). It occurs in all cells but is most important in secretory cells (e.g. cells that produce digestive enzymes) and nerve cells.

The Cytoplasm

Within the plasma membrane is the cytoplasm. It consists of a clear jelly-like fluid called the a) cytosol or intracellular fluid in which b) cell inclusions, c) organelles and d) microfilaments and microtubules are found.

Cytosol : The cytosol consists mainly of water in which various molecules are dissolved or suspended. These molecules include proteins, fats and carbohydrates as well as sodium, potassium, calcium and chloride ions. Many of the reactions that take place in the cell occur in the cytosol.

Cell inclusions : These are large particles of fat, glycogen and melanin that have been produced by the cell. They are often large enough to be seen with the light microscope. For example the cells of adipose tissue (as in the insulating fat layer under the skin) contain fat that takes up most of the cell.

Organelles : Organelles are the "little organs" of the cell - like the heart, kidney and liver are the organs of the body. They are structures with characteristic appearances and specific "jobs" in the cell. Most can not be seen with the light microscope and so it was only when the electron microscope was developed that they were discovered. The main organelles in the cell are the ribosomes, endoplasmic reticulum, mitochondrion, Golgi complex and lysosomes.

Ribosomes

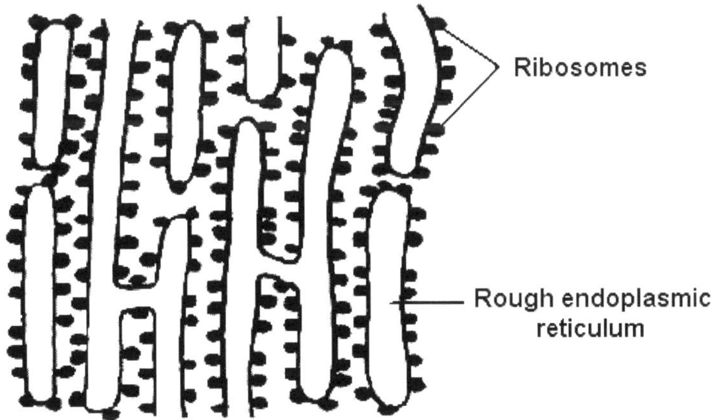

Figure: *Rough endoplasmic reticulum*

Ribosomes are tiny spherical organelles that make proteins by joining amino acids together. Many ribosomes are found free in the cytosol, while others are attached to the rough endoplasmic reticulum.

Endoplasmic Reticulum

The endoplasmic reticulum (ER) is a network of membranes that form channels throughout the cytoplasm from the nucleus to the plasma membrane. Various molecules are made in the ER and transported around the cell in its channels. There are two types of ER: smooth ER and rough ER.

Smooth ER is where the fats in the cell are made and in some cells, where chemicals like alcohol, pesticides and carcinogenic molecules are inactivated.

The Rough ER has ribosomes attached to its surface. The function of the Rough ER is therefore to make proteins that are modified stored and transported by the ER.

Mitochondria

Figure: *A mitochondrion*

Mitochondria (singular mitochondrion) are oval or rod shaped organelles scattered throughout the cytoplasm. They consist of two membranes, the inner one of which is folded to increase its surface area.

Mitochondria are the "power stations" of the cell. They make energy by "burning" food molecules like glucose. This process is called cellular respiration. The reaction requires oxygen and produces carbon dioxide which is a waste product. The process is very complex and takes place in a large number of steps but the overall word equation for cellular respiration is-

Glucose + oxygen = carbon dioxide + water + energy

Note that cellular respiration is different from respiration or breathing. Breathing is the means by which air is drawn into and expelled from the lungs. Breathing is necessary to supply the cells with the oxygen required by the mitochondria and to remove the carbon dioxide produced as a waste product of cellular respiration.

Active cells like muscle, liver, kidney and sperm cells have large numbers of mitochondria.

Golgi Apparatus

Figure: *A Golgi body*

The Golgi bodies in a cell together make up the Golgi apparatus. Golgi bodies are found near the nucleus and consist of flattened membranes stacked on top of each other rather like a pile of plates. The Golgi apparatus modifies and sorts the proteins and fats made by the ER, then surrounds them in a membrane as vesicles so they can be moved to other parts of the cell.

Lysosomes

Lysosomes are large vesicles that contain digestive enzymes. These break down bacteria and other substances that are brought into the cell by phagocytosis or pinocytosis. They also digest worn-out or damaged organelles, the components of which can then be recycled by the cell to make new structures.

Microfilaments And Microtubules

Some cells can move and change shape and organelles and chemicals are moved around the cell. Threadlike structures called microfilaments and microtubules that can contract are responsible for this movement. These structures also form the projections from the plasma membrane known as flagella (singular flagellum) as in the sperm tail, and cilia found lining the respiratory tract and used to remove mucus that has trapped dust particles.

Microtubules also form the pair of cylindrical structures called centrioles found near the nucleus. These help organise the spindle used in cell division.

The Nucleus

The nucleus is the largest structure in a cell and can be seen with the light microscope. It is a spherical or oval body that contains the chromosomes. The nucleus controls the development and activity of the cell. Most cells contain a nucleus although mature red blood cells have lost theirs during development and some muscle cells have several nuclei.

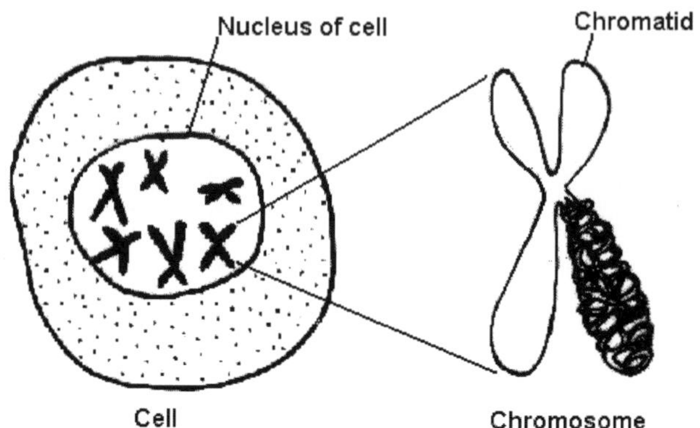

Figure: *A cell with an enlarged chromosome*

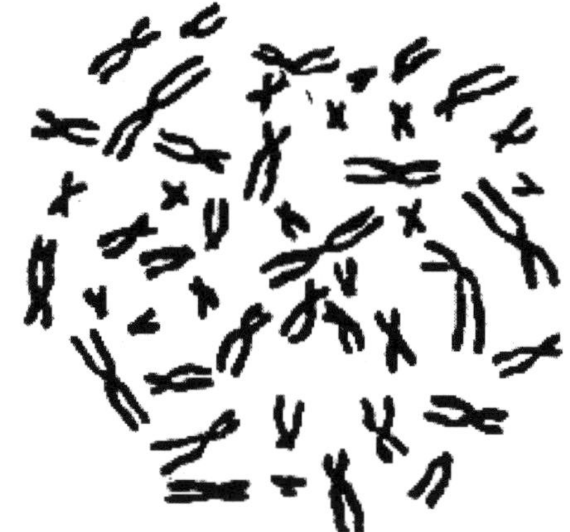

Figure: *A full set of human chromosomes*

A double membrane similar in structure to the plasma membrane surrounds the nucleus (now called the nuclear envelope). Pores in this nuclear membrane allow communication between the nucleus and the cytoplasm. Within the nucleus one or more spherical bodies of darker material can be seen, even with the light microscope. These are called nucleoli and are made of RNA. Their role is to make new ribosomes.

Chromosomes

Inside the nucleus are the chromosomes which:

- contain DNA;
- control the activity of the cell;
- are transmitted from cell to cell when cells divide;

- are passed to a new individual when sex cells fuse together in sexual reproduction.

In cells that are not dividing the chromosomes are very long and thin and appear as dark grainy material. They become visible just before a cell divides when they shorten and thicken and can then be counted. The number of chromosomes in the cells of different species varies but is constant in the cells of any one species (e.g. horses have 64 chromosomes, cats have 38 and humans 46). Chromosomes occur in pairs (i.e. 32 pairs in the horse nucleus and 19 in that of the cat). Members of each pair are identical in length and shape and if you look carefully at diagram, you may be able to see some of the pairs in the human set of chromosomes.

Cell Division

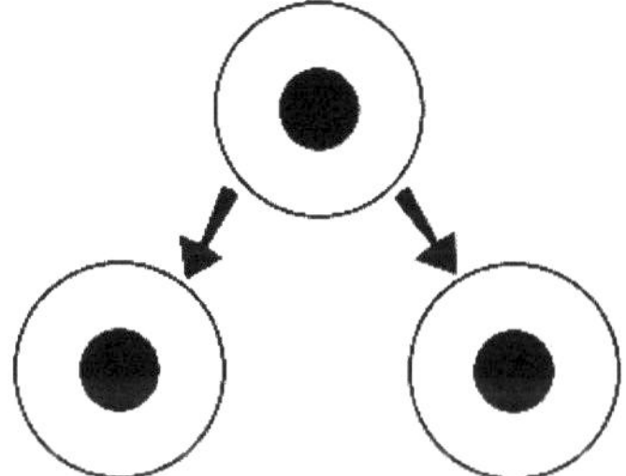

Figure 15: *Division by mitosis results in 2 new cells identical to each other and to parent cell*

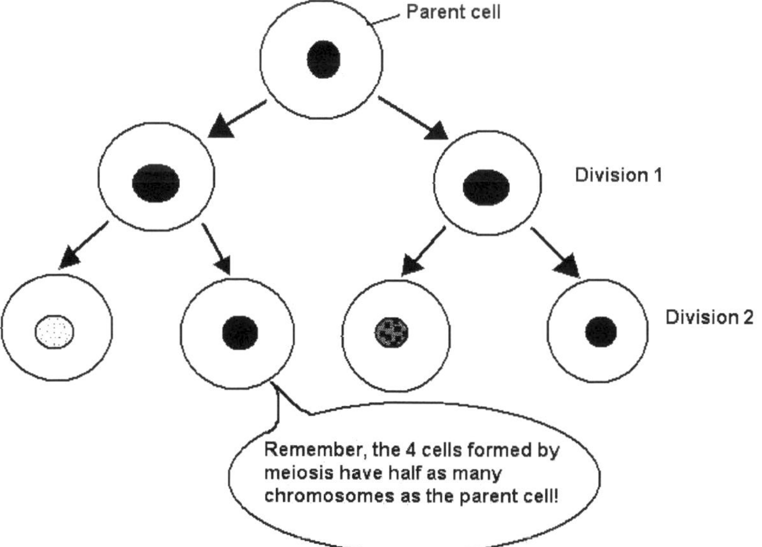

Figure: *Division by meiosis results in 4 new cells that are genetically different to each other*

Cells divide when an animal grows, when its body repairs an injury and when it produces sperm and eggs (or ova). There are two types of cell division: Mitosis and meiosis.

Mitosis: This is the cell division that occurs when an animal grows and when tissues are repaired or replaced. It produces two new cells (daughter cells) each with a full set of chromosomes that are identical to each other and to the parent cell. All the cells of an animal's body therefore contain identical DNA.

Meiosis: This is the cell division that produces the ova and sperm necessary for sexual reproduction. It only occurs in the ovary and testis.

The most important function of meiosis it to halve the number of chromosomes so that when the sperm fertilizes the ovum the normal number is regained. Body cells with the full set of chromosomes are called diploid cells, while gametes (sperm and ova) with half the chromosomes are called haploid cells.

Meiosis is a more complex process than mitosis as it involves two divisions one after the other and the four cells produced are all genetically different from each other and from the parent cell.

This fact that the cells formed by meiosis are all genetically different from each other and from the parent cell can be seen in litters of kittens where all the members of the litter are different from each other as well as being different from the parents although they display characteristics of both.

The Cell as a Factory

To make the function of the parts of the cell easier to understand and remember you can compare them to a factory. For example:

- The nucleus (1) is the managing director of the factory consulting the blueprint (the chromosomes) (2);
- The mitochondria (3) supply the power
- The ribosomes (4) make the products;
- The chloroplasts of plant cells (5) supply the fuel (food)
- The Golgi apparatus (6) packages the products ready for dispatch;
- The ER (7) modifies, stores and transports the products around the factory;
- The plasma membrane is the factory wall and the gates (8);
- The lysosomes dispose of the waste and worn-out machinery.

Body Organisation

The Organisation Of Animal Bodies

Living organisms move, feed, respire (burn food to make energy), grow, sense their environment, excrete and reproduce. These seven characteristics are sometimes summarised by the words "MRS GREN". functions of:

- Movement
- Respiration
- Sensitivity
- Growth
- Reproduction
- Excretion
- Nutrition.

Living organisms are made from cells which are organised into tissues and these are themselves combined to form organs and systems.

Skin cells, muscle cells, skeleton cells and nerve cells, for example. These different types of cells are not just scattered around randomly but similar cells that perform the same function are arranged in groups. These collections of similar cells are known as tissues.

There are four main types of tissues in animals. These are:

- Epithelial tissues that form linings, coverings and glands,
- Connective tissues for transport and support
- Muscle tissues for movement and
- Nervous tissues for carrying messages.

Epithelial Tissues

Epithelium (plural epithelia) is tissue that covers and lines. It covers an organ or lines a tube or space in the body. There are several different types of epithelium, distinguished by the different shapes of the cells and whether they consist of only a single layer of cells or several layers of cells.

Simple Epithelia - with a Single Layer of Cells

Squamous epithelium consists of a single layer of flattened cells that are shaped rather like 'crazy paving'. It is found lining the heart, blood vessels, lung alveoli and body cavities. Its thinness allows molecules to diffuse across readily.

Cuboidal Epithelium

Cuboidal epithelium consists of a single layer of cube shaped cells. It is rare in the body but is found lining kidney tubules. Molecules pass across it by diffusion, osmosis and active transport.

Columnar Epithelium

Columnar epithelium consists of column shaped cells. It is found lining the gut from the stomach to the anus. Digested food products move across it into the blood stream.

Columnar Epithelium with Cilia

Columnar epithelium with cilia on the free surface (also known as the apical side of the cell) lines the respiratory tract, fallopian tubes and uterus. The cilia beat rhythmically to transport particles.

Transitional Epithelium - with a Variable Number of Layers

The cells in transitional epithelium can move over one another allowing it to stretch. It is found in the wall of the bladder.

Stratified Epithelia - with Several Layers of Cells

Epithelia with several layers of cells are found where toughness and resistance to abrasion are needed.

Stratified Squamous Epithelium

Stratified squamous epithelium has many layers of flattened cells. It is found lining the mouth, cervix and vagina. Cells at the base divide and push up the cells above them and cells at the top are worn or pushed off the surface. This type of epithelium protects underlying layers and repairs itself rapidly if damaged.

Keratinised Stratified Squamous Epithelium

Keratinised stratified squamous epithelium has a tough waterproof protein called keratin deposited in the cells. It forms the skin found covering the outer surface of mammals.

Connective Tissues

Blood, bone, tendons, cartilage, fibrous connective tissue and fat (adipose) tissue are all classed as connective tissues. They are tissues that are used for supporting the body or transporting substances around the body. They also consist of three parts: they all have cells suspended in a ground substance or matrix and most have fibres running through it.

Blood

Blood consists of a matrix - plasma, with several types of cells and cell fragments suspended in it. The fibres are only evident in blood that has clotted. Blood will be described in detail in chapter 8.

Lymph

Lymph is similar in composition to blood plasma with various types of white blood cell floating in it. It flows in lymphatic vessels.

Connective Tissue 'Proper'

Connective tissue 'proper' consists of a jelly-like matrix with a dense network of collagen and elastic fibres and various cells embedded in it. There are various different forms of 'proper' connective tissue.

Loose Connective Tissue

Loose connective tissue is a sticky whitish substance that fills the spaces between organs. It is found in the dermis of the skin.

Dense Connective Tissue

Dense connective tissue contains lots of thick fibres and is very strong. It forms tendons, ligaments and heart valves and covers bones and organs like the kidney and liver.

Adipose Tissue

Adipose tissue consists of cells filled with fat. It forms the fatty layer under the dermis of the skin, around the kidneys and heart and the yellow marrow of the bones.

Cartilage

Cartilage is the 'gristle' of the meat. It consists of a tough jelly-like matrix with cells suspended in it. It may contain collagen and elastic fibres. It is a flexible but tough tissue and is found at the ends of bones, in the nose, ear and trachea and between the vertebrae .

Bone

Bone consists of a solid matrix made of calcium salts that give it its hardness. Collagen fibres running through it give it its strength. Bone cells are found in spaces in the matrix. Two types of bone are found in the skeleton namely spongy and compact bone. They differ in the way the cells and matrix are arranged.

Smooth Muscle

Smooth muscle consists of long and slender cells with a central nucleus. It is found in the walls of blood vessels, airways to the lungs and the gut. It changes the size of the blood vessels and helps move food and fluid along. Contraction of smooth muscle fibres occurs without the conscious control of the animal.

Skeletal Muscle

Skeletal muscle (sometimes called striated, striped or voluntary muscle) has striped fibres with alternating light and dark bands. It is attached to bones and is under the voluntary control of the animal.

Cardiac Muscle

Cardiac muscle is found only in the walls of the heart where it produces the 'heart beat'. Cardiac muscle cells are branched cylinders with central nuclei and faint stripes. Each fibre contracts automatically but the heart beat as a whole is controlled by the pacemaker and the involuntary autonomic nervous system.

Nervous Tissues

Nervous tissue forms the nerves, spinal cord and brain. Nerve cells or neurons consist of a cell body and a long thread or axon that carries the nerve impulse. An insulating sheath of fatty material (myelin) usually surrounds the axon. A typical motor neuron that sends messages to muscles to contract.

Vertebrate Bodies

We are so familiar with animals with backbones (i.e. vertebrates) that it seems rather unnecessary to point out that the body is divided into three sections.

There is a well-defined head that contains the brain, the major sense organs and the mouth, a trunk that contains the other organs and a well-developed tail.

Other features of vertebrates may be less apparent. For instance, vertebrates that live on the land have developed a flexible neck that is absent in fish where it would be in the way of the gills and interfere with streamlining. Mammals but not other vertebrates have a sheet of muscle called the diaphragm that divides the trunk into the chest region or thorax and the abdomen.

Body Cavities

In contrast to many primitive animals, vertebrates have spaces or body cavities that contain the body organs. Most vertebrates have a single body cavity but in mammals the diaphragm divides the main cavity into a thoracic and an abdominal cavity.

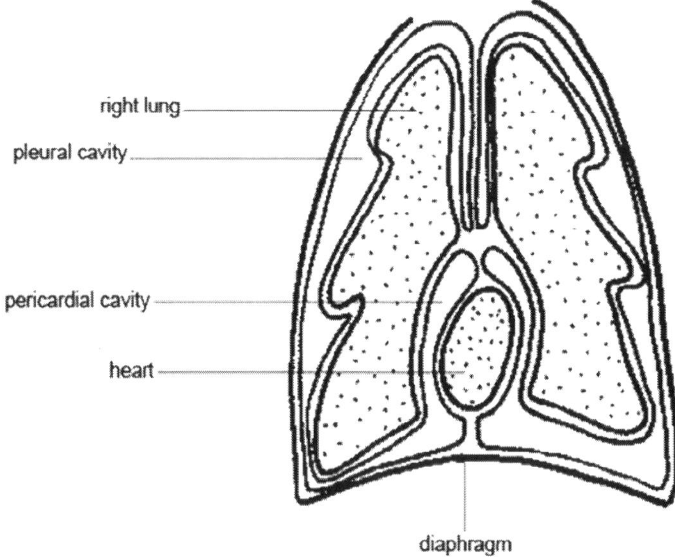

Figure: *The body cavities*

In the thoracic cavity the heart and lungs are surrounded by their own membranes so that cavities are created around the heart - the pericardial cavity, and around the lungs – the pleural cavity.

Organs

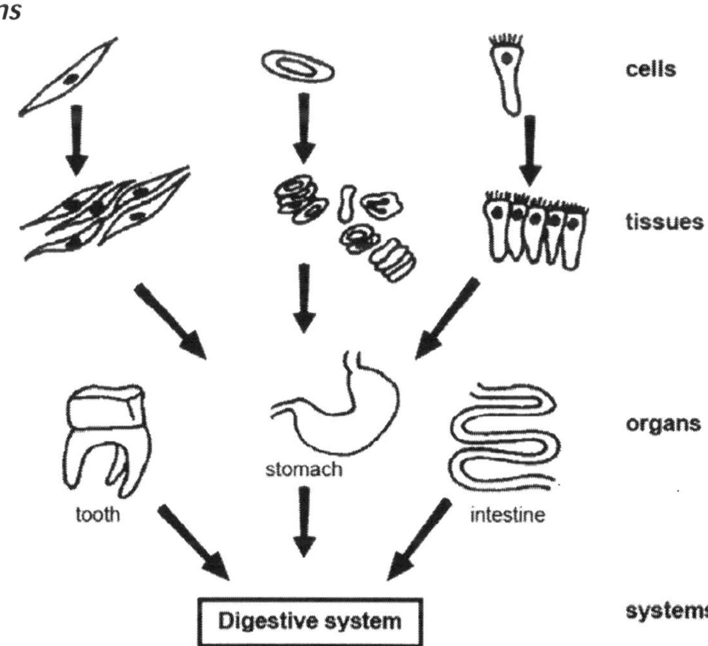

Figure: *Cells, tissues and organs forming the digestive system*

Just as the various parts of the cell work together to perform the cell's functions and a large number of similar cells make up a tissue, so many different tissues can "cooperate" to form an organ that performs a particular function. For example, connective tissues, epithelial tissues, muscle tissue and nervous tissue combine to make the organ that we call the stomach. In turn the stomach combines with other organs like the intestines, liver and pancreas to form the digestive system.

Generalised Plan of The Mammalian Body

At this point it would be a good idea to make yourself familiar with the major organs and their positions in the body of a mammal like the rabbit.

Body Systems

Organs do not work in isolation but function in cooperation with other organs and body structures to bring about the MRS GREN functions necessary to keep an animal alive. For example the stomach can only work in conjunction with the mouth and oesophagus (gullet). These provide it with the food it breaks down and digests. It then needs to pass the food on to the intestines etc. for further digestion and absorption. The organs involved with the taking of food into the body, the digestion and absorption of the food and elimination of waste products are collectively known as the digestive system.

The 11 body Systems

1. *Skin:* The skin covering the body consists of two layers, the epidermis and dermis. Associated with these layers are hairs, feathers, claws, hoofs, glands and sense organs of the skin.
2. *Skeletal System:* This can be divided into the bones of the skeleton and the joints where the bones move over each other.
3. *Muscular System:* The muscles, in conjunction with the skeleton and joints, give the body the ability to move.
4. *Cardiovascular System:* This is also known as the circulatory system. It consists of the heart, the blood vessels and the blood. It transports substances around the body.
5. *Lymphatic System:* This system is responsible for collecting and "cleaning" the fluid that leaks out of the blood vessels. This fluid is then returned to the blood system. The lymphatic system also makes antibodies that protect the body from invasion by bacteria etc. It consists of lymphatic vessels, lymph nodes, the spleen and thymus glands.

6. *Respiratory System:* This is the system involved with bringing oxygen in the air into the body and getting rid of carbon dioxide, which is a waste product of processes that occur in the cell. It is made up of the trachea, bronchi, bronchioles, lungs, diaphragm, ribs and muscles that move the ribs in breathing.
7. *Digestive System:* This is also known as the gastrointestinal system, alimentary system or gut. It consists of the digestive tube and glands like the liver and pancreas that produce digestive secretions. It is concerned with breaking down the large molecules in foods into smaller ones that can be absorbed into the blood and lymph. Waste material is also eliminated by the digestive system.
8. *Urinary System:* This is also known as the renal system. It removes waste products from the blood and is made up of the kidneys, ureters and bladder.
9. *Reproductive System:* This is the system that keeps the species going by making new individuals. It is made up of the ovaries, uterus, vagina and fallopian tubes in the female and the testes with associated glands and ducts in the male.
10. *Nervous System:* This system coordinates the activities of the body and responses to the environment. It consists of the sense organs (eye, ear, semicircular canals, and organs of taste and smell), the nerves, brain and spinal cord.
11. Endocrine System : This is the system that produces chemical messengers or hormones. It consists of various endocrine glands (ductless glands) that include the pituitary, adrenal, thyroid and pineal glands as well as the testes and ovary.

Homeostasis

All the body systems, except the reproductive system, are involved with keeping the conditions inside the animal more or less stable. This is called homeostasis. These constant conditions are essential for the survival and proper functioning of the cells, tissues and organs of the body. The skin, for example, has an important role in keeping the temperature of the body constant. The kidneys keep the concentration of salts in the blood within limits and the islets of Langerhans in the pancreas maintain the correct level of glucose in the blood through the hormone insulin. As long as the various body processes remain within normal limits, the body functions properly and is healthy. Once homeostasis is disturbed disease or death may result.

Directional Terms

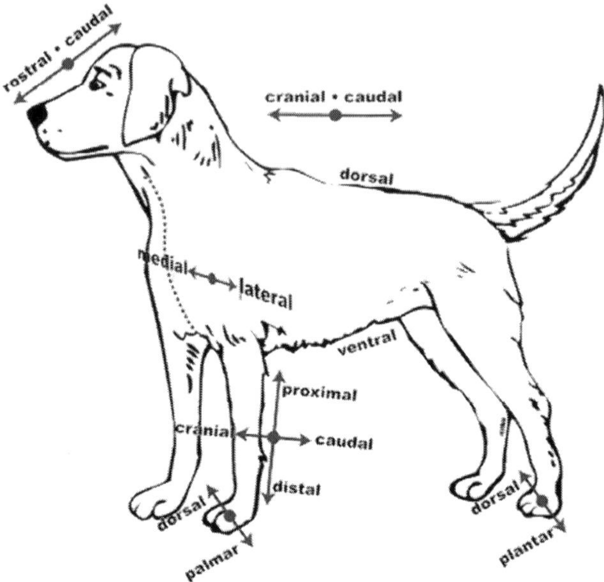

Figure: *The directional terms used with animals*

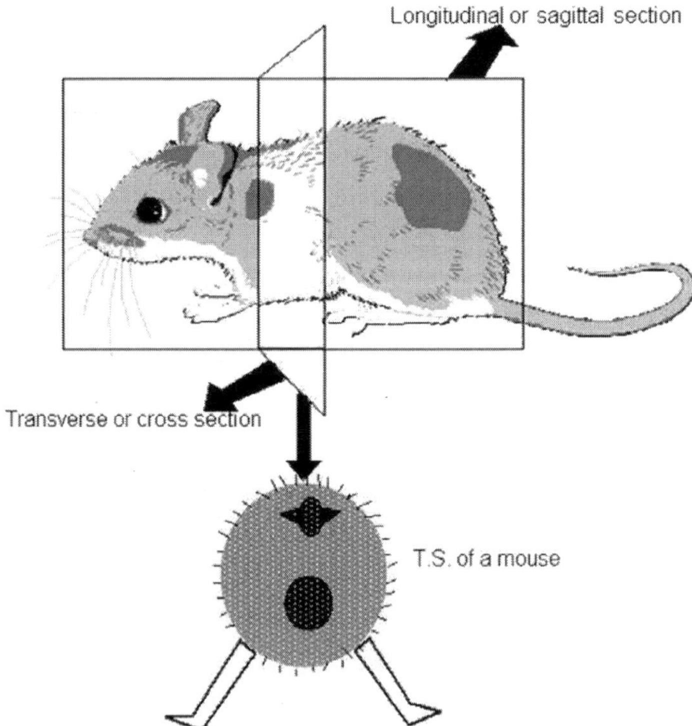

Figure: *Transverse and longitudinal sections of a mouse*

In the following chapters the systems of the body in the list above will be covered one by one. For each one the structure of the organs involved will be described and the way they function will be explained.

In order to describe structures in the body of an animal it is necessary to have a system for describing the position of parts of the body in relation to other parts. For example it may be necessary to describe the position of the liver in relation to the diaphragm, or the heart in relation to the lungs. Certainly if you work further with animals, in a veterinary clinic for example, it will be necessary to be able to accurately describe the position of an injury. The terms used for this are called directional terms.

The most common directional terms are right and left. However, even these are not completely straightforward especially when looking at diagrams of animals. The convention is to show the left side of the animal or organ on the right side of the page. This is the view you would get looking down on an animal lying on its back during surgery or in a post-mortem. Sometimes it is useful to imagine 'getting inside' the animal (so to speak) to check which side is which. The other common and useful directional terms are listed below and shown.

In order to look at the structure of some of the parts or organs of the body it may be necessary to cut them open or even make thin slices of them that they can be examined under the microscope. The direction and position of slices or sections through an animal's body have their own terminology. If an animal or organ is sliced lengthwise this section is called a longitudinal or sagittal section. This is sometimes abbreviated to LS. If the section is sliced crosswise it is called a transverse or cross section. This is sometimes abbreviated to TS or XS.

The Skin

Skin comes in all kinds of textures and forms. There is the dry warty skin of toads and crocodiles, the wet slimy skin of fish and frogs, the hard shell of tortoises and the soft supple skin of snakes and humans. Mammalian skin is covered with hair, that of birds with feathers, and fish and reptiles have scales. Pigment in the skin, hairs or feathers can make the outer surface almost any colour of the rainbow.

As humans, it is often the skin of an animal that gives it its appeal to us or repels us. We love the soft feel of a cat's coat but perhaps can't bear to touch a snake. As the main part of an animal visible to us, the skin can often give us clues to the health of an animal. A healthy animal will have a clean, glowing, flexible skin, while ill health may show itself as an abnormal colour or texture.

Skin is one of the largest organs of the body, making up 6-8% of the total body weight. It consists of two distinct layers. The top layer is called the epidermis and under that is the dermis. The epidermis is the layer that bubbles up when we have a blister and as we know from this experience, it has no blood or nerves in it. The cells at the base of the epidermis continually divide and push the cells above them upwards. As these cells move up they die and become the dry flaky scales that fall off the skin surface. The cells in the epidermis die because a special protein called keratin is deposited in them. Keratin is an extremely important substance for it makes the skin waterproof. Without it land vertebrates like reptiles, birds and mammals would, like frogs, be able to survive only in damp places.

Skin Structures Made of Keratin

Claws, Nails and Hoofs: Reptiles, birds and mammals all have nails or claws on the ends of their toes. They protect the end of the toe and may be used for grasping, grooming, digging or in defence. They are continually worn away and grow continuously from a growth layer at their base.

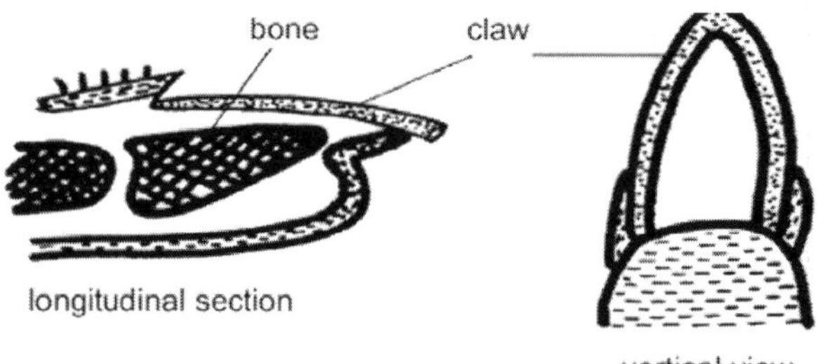

Figure: *A carnivore's claw*

Hoofs are found in sheep, cows, horses etc. otherwise known as ungulate mammals. These are animals that have lost toes in the process of evolution and walk on the "nails" of the remaining toes. The hoof is a cylinder of horny material that surrounds and protects the tip of the toe.

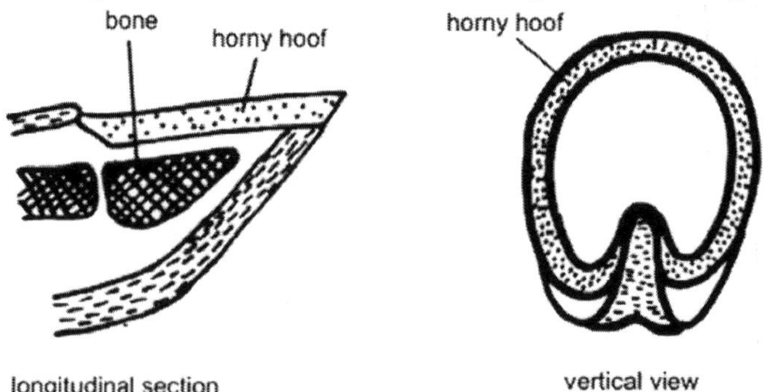

Figure: *A horse's hoof*

Horns And Antlers

True horns are made of keratin and are found in sheep, goats and cattle. They are never branched and, once grown, are never shed. They consist of a core of bone arising in the dermis of the skin and are fused with the skull. The horn itself forms as a hollow cone-shaped sheath around the bone.

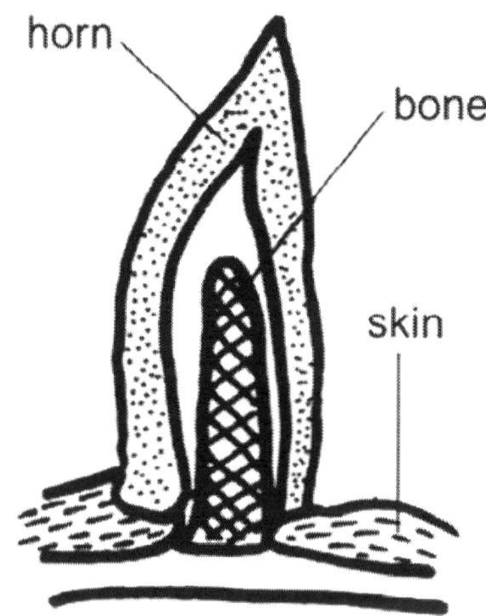

Figure: *A horn*

The antlers of male deer have quite a different structure. They are not formed in the epidermis and do not consist of keratin but are entirely of bone. They are shed each year and are often branched, especially in older animals. When growing they are covered in skin called velvet that forms the bone. Later the velvet is shed to leave the bony antler. The velvet is often removed artificially to be sold in Asia as a traditional medicine.

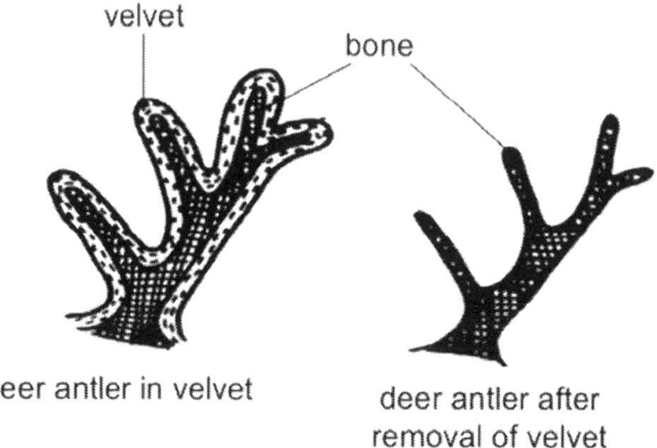

Figure: *A deer antler*

Other animals have projections on their heads that are not true horns either. The horns on the head of giraffes are made of bone covered with skin and hair, and the 'horn' of a rhinoceros is made of modified and fused hair-like structures.

Hair

Hair is also made of keratin and develops in the epidermis. It covers the body of most mammals where it acts as an insulator and helps to regulate the temperature of the body. The colour in hairs is formed from the same pigment, melanin that colours the skin. Coat colour may help camouflage animals and sometimes acts to attract the opposite sex.

Hairs lie in a follicle and grow from a root that is well supplied with blood vessels. The hair itself consists of layers of dead keratin - containing cells and usually lies at a slant in the skin. A small bundle of smooth muscle fibres (the hair erector muscle) is attached to the side of each hair and when this contracts the hair stands on end. This increases the insulating power of the coat and is also used by some animals to make them seem larger when confronted by a foe or a competitor. The whiskers of cats and the spines of hedgehogs are examples of special types of hairs.

Feathers

The lightness and stiffness of keratin is also a key to bird flight. In the form of feathers it provides the large airfoils necessary for flapping and gliding flight. In another form, the light fluffy down feathers,also made of keratin, are some of the best natural insulators known. This superior insulation is necessary to help maintain the high body temperatures of birds.

Countour feathers are large feathers that cover the body, wings and tail. They have an expanded vane that provides the smooth, continuous surface that is required for effective flight. This surface is formed by barbs that extend out from the central shaft. If you look carefully at a feather you can see that on either side of each barb are thousands of barbules that lock together by a complex system of hooks and notches. If this arrangement becomes disrupted, the bird uses its beak to draw the barbs and barbules together again in an action known as preening. Down feathers are the only feathers covering a chick and form the main insulation layer under the contour feathers of the adult. They have no shaft but consist of a spray of simple, slender branches.

Pin feathers have a slender hair-like shaft often with a tiny tuft of barbs on the end. They are found between the other feathers and help tell a bird how its feathers are lying.

Skin Glands

Glands are organs that produce and secrete fluids. They are usually divided into two groups depending upon whether or not they have channels or ducts to carry their products away. Glands with ducts are called exocrine glands and include the glands found in the skin as well as the glands that produce digestive enzymes in the gut. Endocrine glands have no ducts and release their products (hormones) directly into the blood stream. The pituitary and adrenal glands are examples of endocrine glands. Most vertebrates have exocrine glands in the skin that produce a variety of secretions. The slime on the skin of fish and frogs is mucus produced by skin glands and some fish and frogs also produce poison from modified glands.

In fact the skin glands of some frogs produce the most poisonous chemicals known. Reptiles and birds have a dry skin with few glands. The preen gland, situated near the base of the bird's tail, produces oil to help keep the feathers in good condition. Mammals have an array of different skin glands. These include the wax producing, sweat, sebaceous and mammary glands.

Wax producing glands are found in the ears.

Sebaceous glands secrete an oily secretion into the hair follicle. This secretion, known as sebum, keeps the hair supple and helps prevent the growth of bacteria.

Sweat glands consist of a coiled tube and a duct leading onto the skin surface. Their appearance when examined under the microscope inspired one of the first scientists to observe them to call them "fairies' intestines". Sweat contains salt and waste products like urea and the evaporation of sweat on the skin surface is one of the major mechanisms for cooling the body of many mammals. Horses can sweat up to 30 litres of fluid a day during active exercise, but cats and dogs have few sweat glands and must cool themselves by panting. Scent in the sweat of many animals is used to mark territory or attract the opposite sex.

Mammary glands are only present in mammals. They are thought to be modified sebaceous glands and are present in both sexes but are rarely active in males. The number of glands varies from species to species. They open to the surface in well-developed nipples. Milk contains proteins, sugars, fats and salts, although the exact composition varies from one species to another.

The Skin And Sun

A moderate amount of UV in sunlight is necessary for the skin to form vitamin D. This vitamin prevents bone disorders like rickets to which animals reared indoors are susceptible. Excessive exposure to the UV in sunlight can be damaging and the pigment melanin, deposited in cells at the base of the epidermis, helps to protect the underlying layers of the skin from this damage. Melanin also colours the skin and variations in the amount of melanin produces colours from pale yellow to black.

Sunburn And Skin Cancer

Excess exposure to the sun can cause sunburn. This is common in humans, but light skinned animals like cats and pigs can also be sunburned, especially on the ears. Skin cancer can also result from excessive exposure to the sun. As holes in the ozone layer increase exposure to the sun's UV rays, so too does the rate of skin cancer in humans and animals.

The Dermis

The underlying layer of the skin, known as the dermis, is much thicker but much more uniform in structure than the epidermis. It is composed of loose connective tissue with a felted mass of collagen and elastic fibres. It is this part of the skin of cattle and pigs etc. that becomes commercial leather when treated,. The dermis is well supplied with blood vessels, so cuts and burns that penetrate down into the dermis will bleed or cause serious fluid loss. There are also numerous nerve endings and touch receptors in the dermis because, of course, the skin is sensitive to touch, pain and temperature.

When looking at a section of the skin under the microscope you can see hair follicles, sweat and sebaceous glands dipping down into the dermis. However, these structures do not originate in the dermis but are derived from the epidermis.

In the lower levels of the dermis is a layer of fat or adipose tissue. This acts as an energy store and is an excellent insulator especially in mammals like whales with little hair.

The Skin And Temperature Regulation

Vertebrates can be divided into two groups depending on whether or not they control their internal temperature. Amphibia (frogs) and reptiles are said to be cold blooded" (poikilothermic) because their body temperature approximately follows that of the environment. Birds and mammals are said to be warm blooded (homoiothermic) because they can maintain a roughly constant body temperature despite changes in the temperature of the environment. Heat is produced by the biochemical reactions of the body (especially in the liver) and by muscle contraction. Most of the heat lost from the body occurs via the skin. It is therefore not surprising that many of the mechanisms for controlling the temperature of the body operate here.

Reduction of Heat Loss

When an animal is in a cold environment and needs to reduce heat loss the erector muscles contract causing the hair or feathers to rise up and increase the layer of insulating air trapped by them. Heat loss from the skin surface can also be reduced by the contraction of the abundant blood vessels that lie in the dermis. This takes blood flow to deeper levels, so reducing heat loss and causing pale skin. Shivering caused by twitching muscles produces heat that also helps raise the body temperature.

Increase of Heat Loss

There are two main mechanisms used by animals to increase the amount of heat lost from the skin when they are in a hot environment or high levels of activity are increasing internal heat production. The first is the expansion of the blood vessels in the dermis so blood flows near the skin surface and heat loss to the environment can take place. The second is by the production of sweat from the sweat glands. The evaporation of this liquid on the skin surface produces a cooling effect. The mechanisms for regulating body temperature are under the control of a small region of the brain called the hypothalamus. This acts like a thermostat.

Heat Loss and Body Size

The amount of heat that can be lost from the surface of the body is related to the area of skin an animal has in relation to the total volume of its body. Small animals like mice have a very large skin area compared to their total volume. This means they tend to loose large amounts of heat and have difficulty keeping warm in cold weather. They may need to keep active just to maintain their body temperature or may hibernate to avoid the problem. Large animals like elephants have the opposite problem. They have only a relatively small skin area in relation to their total volume and may have trouble keeping cool. This is one reason that these large animals tend to have sparse coverings of hair.

Anatomy and Physiology of Animals/The Skeleton

Fish, frogs, reptiles, birds and mammals are called vertebrates, a name that comes from the bony column of vertebrae (the spine) that supports the body and head. The rest of the skeleton of all these animals (except the fish) also has the same basic design with a skull that houses and protects the brain and sense organs and ribs that protect the heart and lungs and, in mammals, make breathing possible. Each of the four limbs is made to the same basic pattern. It is joined to the spine by means of a flat, broad bone called a girdle and consists of one long upper bone, two long lower bones, several smaller bones in the wrist or ankle and five digits.

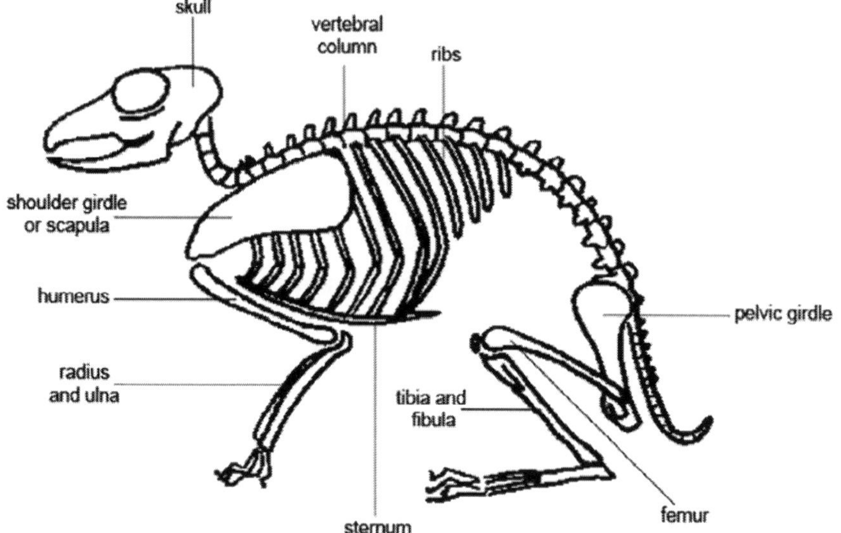

Figure: *The mammalian skeleton*

The Vertebral Column

The vertebral column consists of a series of bones called vertebrae linked together to form a flexible column with the skull at one end and the tail at the other. Each vertebra consists of a ring of bone with spines (spinous process) protruding dorsally from it. The spinal cord passes through the hole in the middle and muscles attach to the spines making movement of the body possible.

The shape and size of the vertebrae of mammals vary from the neck to the tail. In the neck there are cervical vertebrae with the two top ones, the atlas and axis, being specialised to support the head and allow it to nod "Yes" and shake "No". Thoracic vertebrae in the chest region have special surfaces against which the ribs move during breathing. Grazing animals like cows and giraffes that have to support weighty heads on long necks have extra large spines on their cervical and thoracic vertebrae for muscles to attach to. Lumbar vertebrae in the loin region are usually large strong vertebrae with prominent spines for the attachment of the large muscles of the lower back. The sacral vertebrae are usually fused into one solid bone called the sacrum that sits within the pelvic girdle. Finally there are a variable number of small bones in the tail called the coccygeal vertebrae.

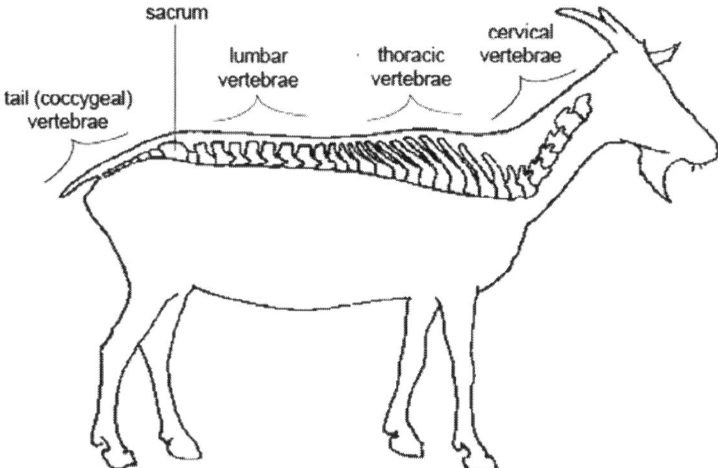

Figure: *The regions of the vertebral column dik*

The Skull

The skull of mammals consists of 30 separate bones that grow together during development to form a solid case protecting the brain and sense organs. The "box "enclosing and protecting the brain is called the cranium. The bony wall of the cranium encloses the middle and inner ears, protects the organs of smell in the nasal cavity and the eyes in sockets known as orbits. The teeth are inserted into the upper and lower jaws. The lower jaw is known as the mandible. It forms a joint with the skull moved by strong muscles that allow an animal to chew. At the front of the skull is the nasal cavity, separated from the mouth by a plate of bone called the palate. Behind the nasal cavity and connecting with it are the sinuses. These are air spaces in the bones of the skull which help keep the skull as light as possible. At the base of the cranium is the foramen magnum, translated as "big hole", through which the spinal cord passes. On either side of this are two small, smooth rounded knobs or condyles that articulate (move against) the first or Atlas vertebra.

The Rib

Paired ribs are attached to each thoracic vertebra against which they move in breathing. Each rib is attached ventrally either to the sternum or to the rib in front by cartilage to form the rib cage that protects the heart and lungs. In dogs one pair of ribs is not attached ventrally at all. They are called floating ribs. Birds have a large expanded sternum called the keel to which the flight muscles (the 'breast" meat of a roast chicken) are attached.

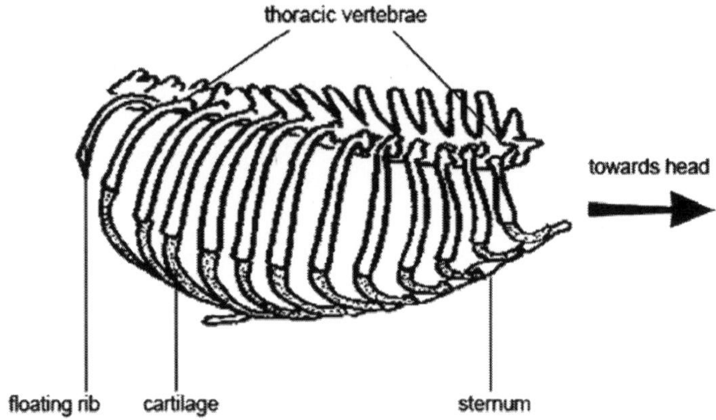

Figure: *The rib*

The Forelimb

The forelimb consists of: Humerus, radius and ulna, carpals, metacarpals, digits or phalanges. The top of the humerus moves against (articulates with) the scapula at the shoulder joint. By changing the number, size and shape of the various bones, fore limbs have evolved to fit different ways of life. They have become wings for flying in birds and bats, flippers for swimming in whales, seals and porpoises, fast and efficient limbs for running in horses and arms and hands for holding and manipulating in primates.

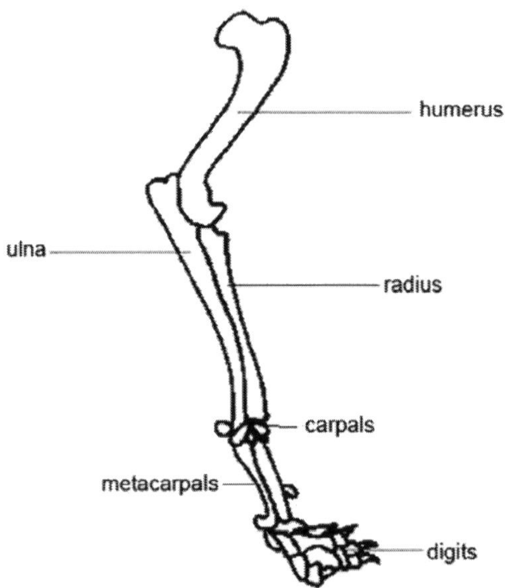

Figure: *Forelimb of a dog*

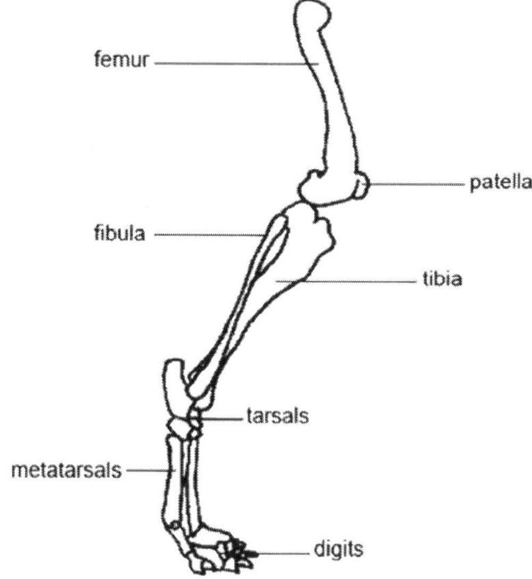

Figure: *Hindlimb of a dog*

The Hind Limb

The hind limbs have a similar basic pattern to the forelimb. They consist of: femur, tibia and fibula, tarsals, metatarsals, digits or phalanges. The top of the femur moves against (articulates with) the pelvis at the hip joint.

The Long pastern or proximal phalanx is P1, not P3. The distal phalanx or coffin bone (called hoof there) is the 3rd phalanges. The patella or kneecap is embedded in a large tendon in front of the knee. It seems to smooth the movements of the knee. The legs of the horse are highly adapted to give it great galloping speed over long distances. The bones of the leg, wrist and foot are greatly elongated and the hooves are actually the tips of the third fingers and toes, the other digits having been lost or reduced.

The Girdles

The girdles pass on the "push" produced by the limbs to the body. The shoulder girdle or scapula is a triangle of bone surrounded by the muscles of the back but not connected directly to the spine. This arrangement helps it to cushion the body when landing after a leap and gives the forelimbs the flexibility to manipulate food or strike at prey. Animals that use their forelimbs for grasping, burrowing or climbing have a well-developed clavicle or collar bone. This connects the shoulder girdle to the sternum. Animals like sheep, horses and cows that use their forelimbs only for supporting the body and locomotion have no clavicle. The pelvic girdle or hipbone attaches the sacrum and the hind legs. It transmits the force of the leg-thrust in walking or jumping directly to the spine.

Categories of Bones

People who study skeletons place the different bones of the skeleton into groups according to their shape or the way in which they develop. Thus we have long bones like the femur, radius and finger bones, short bones like the ones of the wrist and ankle, irregular bones like the vertebrae and flat bones like the shoulder blade and bones of the skull. Finally there are bones that develop in tissue separated from the main skeleton. These include sesamoid bones which include bones like the patella or kneecap that develop in tendons and visceral bones that develop in the soft tissue of the penis of the dog and the cow's heart.

Bird Skeletons

Although the skeleton of birds is made up of the same bones as that of mammals, many are highly adapted for flight. The most noticeable difference is that the bones of the forelimbs are elongated to act as wings. The large flight muscles make up as much as 1/5th of the body weight and are attached to an extension of the sternum called the keel. The vertebrae of the lower back are fused to provide the rigidity needed to produce flying movements. There are also many adaptations to reduce the weight of the skeleton. For instance birds have a beak rather than teeth and many of the bones are hollow.

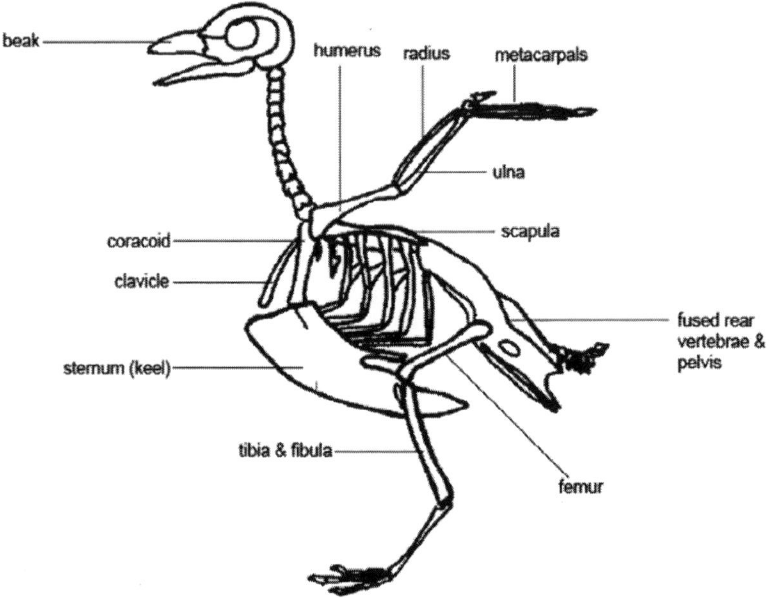

Figure: *A bird's skeleton*

The Structure of Long Bones

A long bone consists of a central portion or shaft and two ends called epiphyses. Long bones move against or articulate with other bones at joints and their ends have flattened surfaces and rounded protuberances (condyles) to make this possible. If you carefully examine a long bone you may also see raised or rough surfaces. This is where the muscles that move the bones are attached. You will also see holes (a hole is called a foramen) in the bone. Blood vessels and nerves pass into the bone through these. You may also be able to see a fine line at each end of the bone. This is called the growth plate or epiphyseal line and marks the place where increase in length of the bone occurred.

If you cut a long bone lengthways you will see it consists of a hollow cylinder. The outer shell is covered by a tough fibrous sheath to which the tendons are attached. Under this is a layer of hard, dense compact bone.

This gives the bone its strength. The central cavity contains fatty yellow marrow, an important energy store for the body, and the ends are made from honeycomb-like bony material called spongy bone. Spongy bone contains red marrow where red blood cells are made.

Compact Bone

Compact bone is not the lifeless material it may appear at first glance. It is a living dynamic tissue with blood vessels, nerves and living cells that continually rebuild and reshape the bone structure as a result of the stresses, bends and breaks it experiences. Compact bone is composed of microscopic hollow cylinders that run parallel to each other along the length of the bone.

Each of these cylinders is called a Haversian system. Blood vessels and nerves run along the central canal of each Haversian system. Each system consists of concentric rings of bone material (the matrix) with minute spaces in it that hold the bone cells. The hard matrix contains crystals of calcium phosphate, calcium carbonate and magnesium salts with collagen fibres that make the bone stronger and somewhat flexible. Tiny canals connect the cells with each other and their blood supply.

Spongy Bone

Spongy bone gives bones lightness with strength. It consists of an irregular lattice that looks just like an old fashioned loofah sponge. It is found on the ends of long bones and makes up most of the bone tissue of the limb girdles, ribs, sternum, vertebrae and skull. The spaces contain red marrow, which is where red blood cells are made and stored.

Bone Growth

The skeleton starts off in the foetus as either cartilage or fibrous connective tissue. Before birth and, sometimes for years after it, the cartilage is gradually replaced by bone. The long bones increase in length at the ends at an area known as the epiphyseal plate where new cartilage is laid down and then gradually converted to bone. When an animal is mature, bone growth ceases and the epiphyseal plate converts into a fine epiphyseal line.

Broken Bones

A fracture or break dramatically demonstrates the dynamic nature of bone. Soon after the break occurs blood pours into the site and cartilage is deposited. This starts to connect the broken ends together. Later spongy bone replaces the cartilage, which is itself replaced by compact bone. Partial healing to the point where some weight can be put on the bone can take place in 6 weeks but complete healing may take 3-4 months.

Joints

Joints are the structures in the skeleton where 2 or more bones meet. There are several different types of joints. Some are immovable once the animal has reached maturity. Examples of these are those between the bones of the skull and the midline joint of the pelvic girdle. Some are slightly moveable like the joints between the vertebrae but most joints allow free movement and have a typical structure with a fluid filled cavity separating the articulating surfaces (surfaces that move against each other) of the two bones.

This kind of joint is called a synovial joint. The joint is held together by bundles of white fibrous tissue called ligaments and a fibrous capsule encloses the joint. The inner layers of this capsule secrete the synovial fluid that acts as a lubricant. The articulating surfaces of the bones are covered with cartilage that also reduces friction and some joints, e.g. the knee, have a pad of cartilage between the surfaces that articulate with each other.

The shape of the articulating bones in a joint and the arrangement of ligaments determine the kind of movement made by the joint. Some joints only allow a to and from gliding movement e.g. between the ankle and wrist bones; the joints at the elbow, knee and fingers are hinge joints and allow movement in two dimensions and the axis vertebra pivots on the atlas vertebra. Ball and socket joints, like those at the shoulder and hip, allow the greatest range of movement.

Common Names of Joints

Some joints in animals are given common names that tend to be confusing. For example:

1. The joint between the femur and the tibia on the hind leg is our knee but the stifle in animals.
2. Our ankle joint (between the tarsals and metatarsals) is the hock in animals
3. Our knuckle joint (between the metacarpals or metatarsals and the phalanges) is the fetlock in the horse.
4. The "knee" on the horse is equivalent to our wrist (ie on the front limb between the radius and metacarpals).

Locomotion

Different animals place different parts of the foot or forelimb on the ground when walking or running.

Humans and bears put the whole surface of the foot on the ground when they walk. This is known as plantigrade locomotion. Dogs and cats walk on their toes (digitigrade locomotion) while horses and pigs walk on their "toenails" or hoofs. This is called unguligrade locomotion.

1. Plantigrade locomotion (on the "palms of the hand) as in humans and bears
2. Digitigrade locomotion (on the "fingers") as in cats and dogs
3. Unguligrade locomotion (on the "fingernails") as in horses.

Muscles

Muscles make up the bulk of an animal's body and account for about half its weight. The meat on the chop or roast is muscle and is composed mainly of protein. The cells that make up muscle tissue are elongated and able to contract to a half or even a third of their length when at rest. There are three different kinds of muscle; smooth, cardiac and skeletal muscle.

Smooth Muscle

Smooth or Involuntary muscle carries out the unconscious routine tasks of the body such as moving food down the digestive system, keeping the eyes in focus and adjusting the diametre of blood vessels. The individual cells are spindle-shaped, being fatter in the middle and tapering off towards the ends with a nucleus in the centre of the cell. They are usually found in sheets and are stimulated by the non-conscious or autonomic nervous system as well as by hormones.

Cardiac Muscle

Cardiac muscle is only found in the wall of the heart. It is composed of branching fibres that form a three-dimensional network. When examined under the microscope, a central nucleus and faint stripes or striations can be seen in the cells.

Cardiac muscle cells contract spontaneously and rhythmically without outside stimulation but the pacemaker coordinates the heart beat. Nerves and hormones modify this rhythm.

Skeletal Muscle

Skeletal muscle is the muscle that is attached to and moves the skeleton, and is under voluntary control. It is composed of elongated cells or fibres lying parallel to each other. Each cell is unusual in that it has several nuclei and when examined under the microscope appears striped or striated.

This appearance gives the muscle its names of striped or striated muscle. Each cell of striated muscle contains hundreds, or even thousands, of microscopic fibres each one with its own striped appearance. The stripes are formed by two different sorts of protein that slide over each other making the cell contract.

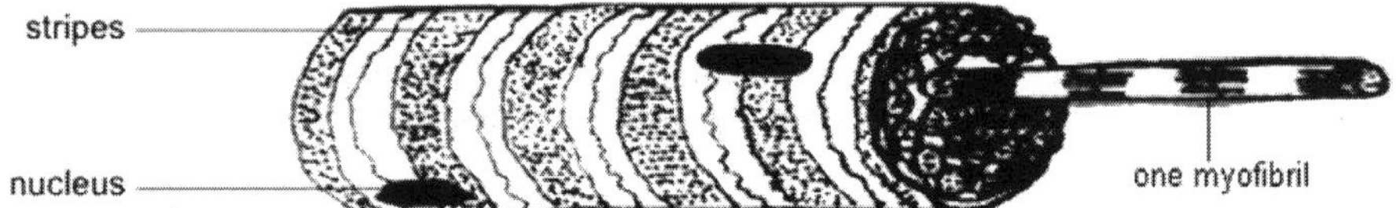

***Figure:** A striped muscle cell*

Muscle Contraction

Muscle contraction requires energy and muscle cells have numerous mitochondria.

However, only about 15% of the energy released by the mitochondria is used to fuel muscle contraction. The rest is released as heat. This is why exercise increases body temperature and makes animals sweat or pant to rid themselves of this heat.

What we refer to as a muscle is made up of groups of muscle fibres surrounded by connective tissue. The connective tissue sheaths join together at the ends of the muscle to form tough white bands of fibre called tendons.

These attach the muscles to the bones. Tendons are similar in structure to the ligaments that attach bones together across a joint.

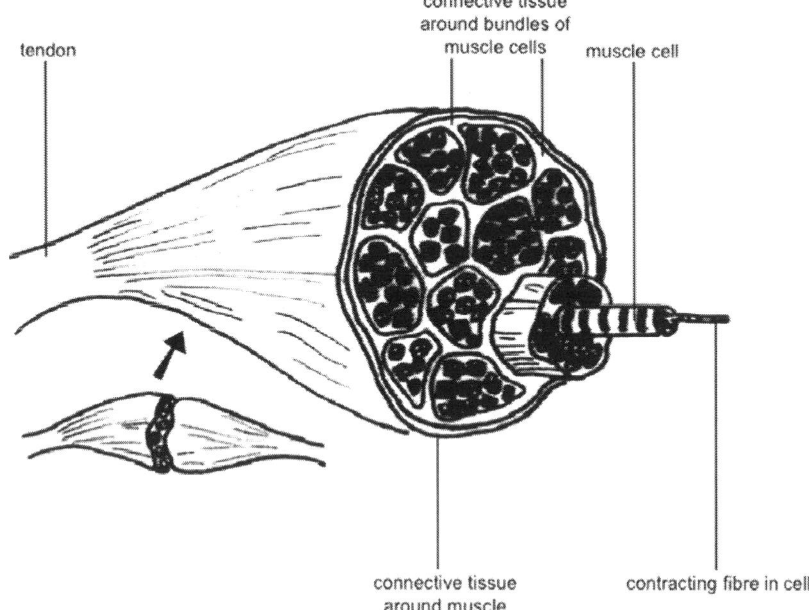

Figure: *The structure of a muscle*

Structure of a Muscle

A single muscle is fat in the middle and tapers towards the ends. The middle part, which gets fatter when the muscle contracts, is called the belly of the muscle. If you contract your biceps muscle in your upper arm you may feel it getting fatter in the middle. You may also notice that the biceps is attached at its top end to bones in your shoulder while at the bottom it is attached to bones in your lower arm. Notice that the bones at only one end move when you contract the biceps. This end of the muscle is called the insertion. The other end of the muscle, the origin, is attached to the bone that moves the least.

Antagonistic Muscles

Skeletal muscles usually work in pairs. When one contracts the other relaxes and vice versa. Pairs of muscles that work like this are called antagonistic muscles. For example the muscles in the upper forearm are the biceps and triceps. Together they bend the elbow. When the biceps contracts (and the triceps relaxes) the lower forearm is raised and the angle of the joint is reduced. This kind of movement is called flexion. When the triceps is contracted (and the biceps relaxes), the angle of the elbow increases. The term for this movement is extension.

When you or animals contract skeletal muscle it is a voluntary action. For example, you make a conscious decision to walk across the room, raise the spoon to your mouth or smile. There is however, another way in which contraction of muscles attached to the skeleton happens that is not under voluntary control. This is during a reflex action, such as jerking your hand away from the hot stove you have touched by accident.

Cardiovascular System/Blood

Blood is a unique fluid containing cells that is pumped by the heart around the body of animals in a system of pipes known as the circulatory system. It carries oxygen and nutrients to the cells of the body and removes waste products like carbon dioxide from them. Blood is also important for keeping conditions in the body constant, in other words for maintaining homeostasis. It helps keep the acidity or pH stable and helps maintain a constant temperature in the body. Blood also has an important role in defending the body against disease.

A simple way to find out what is in blood is to remove a small amount from an animal and place it in a tube with a substance that prevents it from clotting (an anticoagulant). If you leave the tube to stand for a few hours you will find that it settles out into two layers. The top layer consists of a light yellow fluid, the plasma, and the bottom layer consists of red blood cells (RBCs). If you look very carefully you can also see a thin beige-coloured layer in between these two layers. This consists of the white blood cells (WBCs).

The above procedure is usually done more rapidly by placing the blood sample in a centrifuge for a few minutes. This machine acts like a super spin drier rotating about 10,000 times a minute and packing the heavier particles (red blood cells) at the bottom of the tube. The sample that results is called the packed cell volume (P.C.V.) or haematocrit. It is a very useful measurement of the concentration of red blood cells in the blood. For most animals the packed cell volume is in the range 30-45%. If it is lower than this it means that the concentration of red blood cells is low and the animal is anaemic. If the reading is above this range it may mean the animal is dehydrated. Animals that live at high altitudes also have high P.C.V.s to compensate for the low oxygen concentration there.

Plasma

Plasma consists of water (91%) in which many substances are dissolved. These dissolved substances include:

- salts (or electrolytes)
- proteins
- nutrients

- waste products
- dissolved gases (mainly carbon dioxide)
- and other chemicals like hormones.

Salts in Plasma: Salts in the plasma are in the form of ions or electrolytes which include sodium, potassium, calcium, chloride, phosphate and bicarbonate. Plasma transports these ions to where they are needed e.g. calcium required by the bones, they also help keep the osmotic pressure and acid-base balance (pH) of the blood within the required levels.

Blood Proteins

The proteins in the blood plasma are large molecules with important functions. Some contribute to the osmotic pressure and the viscosity (thickness) of the blood, and so help keep the blood volume and pressure stable. Others act as antibodies that attack bacteria and viruses, and yet others are important in blood clotting. Nutrients that are absorbed from the gut and transported to the cells in the plasma include amino acids, glucose, fatty acids and vitamins. Waste products include urea from the breakdown of proteins.

Red Blood Cells

Red blood cells are also known as RBCs or erythrocytes. They are what make blood red. When you look at a blood smear through a microscope, as you will in one of the practical classes, you will see that RBCs are by far the most common cells in the blood. If you focus on an individual RBC, you will see that they are shaped like discs or doughnuts with a thin central portion surrounded by a fatter margin. This shape has all sorts of advantages, one being that enables the cells to fold up and pass along the narrowest blood capillaries.

Front and side view of a red blood cell

Red blood cell cut in half

Red blood cells as they appear in a blood clot

Figure: *Red blood cells or erythrocytes*

The mature RBCs of mammals have neither nucleus nor other organelles and can be thought of as sacks of haemoglobin. Haemoglobin is a red coloured protein containing iron, which joins with oxygen so the blood can transport it to body cells. RBCs are made continuously in the bone marrow and live about 120 days. They are then destroyed in the liver and spleen and the molecules they are made from recycled to make new RBCs. Anaemia results if the rate at which RBCs are destroyed exceeds the rate at which RBC'c are produced.

Note that if you happen to look at bird's, reptiles, frogs or fishes blood down the microscope you will see that these vertebrates all have RBCs with a central nucleus.

White Blood Cells

White blood cells or leucocytes are far less numerous than red blood cells. In fact there is only about one white cell for every 1000 red blood cells. Rather than being white, they are actually colourless as they contain no haemoglobin although unlike RBCs they do have a nucleus. If you make a blood smear and look at it under the microscope it is difficult to see the white blood cells at all. To make them visible you need to stain them with special dyes or stains. There are a variety of stains that can be used, but most dye the nucleus a dark purple or pink colour. The stains may also show up the granules present in the cytoplasm of some white blood cells. White blood cells are divided into two major groups depending on the shape of the nucleus and whether or not there are granules in the cytoplasm.

1. Granulocytes or polymorphonuclear leucocytes ("polymorphs" or "polys") have granules in the cytoplasm and a purple lobed nucleus. The most common (neutrophils) can squeeze out of capillaries and are involved in engulfing and destroying foreign invaders like bacteria. Some (eosinophils) combat allergies and increase in numbers during parasitic worm infections. Others (basophils) produce heparin that prevents the blood from clotting.
2. Agranulocytes or monomorphonuclear leucocytes have a large unlobed nucleus and no granules in the cytoplasm. There are two types of agranulocytes. The most numerous are lymphocytes that are concerned with immune responses. The second type is the monocyte that is the largest blood cell and is involved in engulfing bacteria etc. by phagocytosis.

Platelets

As well as red and white blood cells, the blood also contains small irregular shaped fragments of cells known as platelets. They are involved in the clotting of the blood.

Transport of Oxygen

The purpose of the haemoglobin in red blood cells is to carry oxygen from the lungs to the tissues. In fact it allows the blood to carry about 25 times more oxygen than it would be able to without any haemoglobin.

When oxygen concentrations are high, as in the blood capillaries in the lungs, haemoglobin combines with oxygen to form a compound called oxyhaemoglobin. This compound is bright red and makes the oxygenated blood that spurts from a damaged artery its characteristic bright red colour. When the blood reaches the tissues where the oxygen concentrations are low, the oxygen separates from the haemoglobin and diffuses into the tissues. The haemoglobin in most veins has given up its oxygen and the blood is called deoxygenated blood. It is a purple-red colour.

Carbon Monoxide Poisoning

Carbon monoxide is a colourless, odourless gas found in car exhaust fumes and tobacco smoke. It combines with haemoglobin just like oxygen but does not let go. This means the haemoglobin molecules are not available to carry oxygen to the tissues and the animal or human suffocates. Carbon monoxide poisoning is often fatal but can be treated by giving the patient pure oxygen that slowly replaces the carbon monoxide.

Transport of Carbon Dioxide

Carbon dioxide is a waste gas produced by cells. It diffuses into the blood capillaries where it is carried to the lungs in the blood. Most is carried in the plasma as bicarbonate ions but a small amount is dissolved directly in the plasma and some combines with haemoglobin.

Transport of Other Substances

The blood carries water to the cells and organs as well as soluble food substances (sugars, amino acids, fatty acids and vitamins) and hormones dissolved in the plasma. These are delivered to the cells via the tissue fluid that surrounds them. Blood also picks up the waste products like carbon dioxide and urea from the cells and is important in distributing the heat produced in the liver and muscles all over the body.

Blood Clotting

The mechanism that causes the blood to clot is easily seen when you or your animals are injured. However, minor injuries occur all the time in areas that experience wear and tear like the intestine, the lungs and the skin. Without the clotting mechanism, animals would quickly bleed to death from minor injury and internal haemorrhage. This is what happens in animals and people with clotting disorders like haemophilia, as well as animals that are poisoned with rat poisons like warfarin.

Platelets are important in blood clotting. When blood vessels are damaged, substances released cause the blood platelets to disintegrate. This stimulates a complex chain of reactions, which causes the protein fibrinogen to be converted to fibrin. Fibrin forms a dense fibrous network over the wound preventing the escape of further blood. Calcium and vitamin K are essential for the clotting process and any deficiency of these may also lead to clotting problems.

Serum and Plasma

When blood clots it separates into the clot that contains most of the cells and platelets leaving behind a straw coloured fluid. This fluid is called serum. It looks just like plasma and is similar in composition except for one big difference. It doesn't contain fibrinogen, the protein that forms the clot.

Anticoagulants

Anticoagulants are substances that interfere with the clotting process. When blood is collected for transfusion or testing it is often important to prevent it clotting and there are a number of different anticoagulants you can use for this. Tubes containing the different anticoagulants are coded with different colours for easy recognition.

1. Heparin (colour code - green) is a natural anticoagulant produced by the white blood cells but it is also used routinely in the laboratory with samples to be tested for heavy metals like lead.
2. EDTA (colour code – lavender) is used for routine blood counts.
3. Fluoroxylate (colour code – grey) is used for biochemical tests for glucose.
4. Citrate (colour code – light blue) is used for the storage of large quantities of blood, such as used in transfusions.

Haemolysis

Haemolysis is the breakdown of the plasma membrane of red blood cells to release the haemoglobin. We have already met this process when discussing osmosis, for haemolysis often occurs when red blood cells are placed in a hypotonic solution and water flows in through the semi permeable plasma membrane to swell and eventually burst the cell. It is therefore important when collecting blood from an animal to make sure there is no water in the syringe or tube. Too much movement due to shaking the tube or sucking up the blood too vigorously can also break down the plasma membrane and cause haemolysis.

Blood Groups

If you have given blood recently you may know your blood group. It may be blood group O, A or B or even AB, the rarest group. Blood groups are the result of different molecules called antigens on the outside of red blood cells. These cause antibodies to be formed that attack viruses and bacteria. Knowledge of a person's blood group is important when giving transfusions because if blood of another incompatible blood group is given to a patient the red blood cells stick together and block the blood vessels and may lead to death.

Blood groups also exist in many animals. There are three blood groups in cats and great care has to be taken that the groups are compatible when transfusing exotic breeds. The situation is slightly different in dogs. They have a number of blood groups but there is usually no problem with the first blood transfusion a dog receives. However, this first transfusion sensitises the immune system so that a problem may arise with the second and subsequent transfusions.

Haemolysis can occur in the living animal when it is exposed to various poisons and toxins. This may happen when, for example, it eats a poisonous plant, is bitten by a snake or infected with bacteria that destroy red blood cells (haemolytic bacteria).

Blood Volume

Blood accounts for between 6-10% of the body weight of animals, varying with the species and the stage of life. Animals can not tolerate losses of greater than 3% of the total volume when the condition known as shock occurs.

Summary Blood

- The main functions of blood are transport of oxygen, food, waste products etc., the maintenance of homeostasis and defending the body from disease.
- Blood consists of fluid, plasma, in which red and white blood cells are suspended. The blood cells typically make up 30-45% of the blood volume.
- Plasma consists of water containing dissolved substances like proteins, nutrients and carbon dioxide.
- Red Blood Cells contain haemoglobin to transport oxygen.
- White Blood Cells defend the body from invasion. There are 2 kinds:
  - Granular white cells include neutrophils, basophils and eosinophils. Neutrophils which destroy bacteria are the most numerous. Eosinophils are involved with allergies and parasitic infections.
  - Non-granular white cells include lymphocytes that produce antibodies to attach bacteria and viruses and monocytes that engulf and destroy bacteria and viruses.
- Platelets are involved in blood clotting.

Chapter 7

Emotion in Animals

There is no scientific consensus on emotion in animals, that is, what emotions certain species of animals, including humans, feel. The debate concerns primarily mammals and birds, although emotions have also been postulated for other vertebrates and even for some invertebrates.

Animal lovers, scientists, philosophers, and others who interact with animals, have suggested answers but the core question has proven difficult to answer since animals cannot speak of their experience. Society recognises that animals can feel pain, as demonstrated by the criminalisation of animal cruelty. Animal expressions of apparent pleasure are ambiguous as to whether this is emotion, or simply innate responses, perhaps for approval or other hard-wired cues. The ambiguity is a source of controversy as there is no certainty which views, if any, reflect reality. That said, extreme behaviourists would say that human "feeling" is also merely a hard-wired response to external stimuli. In recent years, research has become available which expands prior understandings of animal language, cognition and tool use, and even sexuality. Emotions arise in the mammalian brain, or the limbic system, which human beings share in common with other mammals as well as many other species.

Evidence

While humans have had differing views of animal emotion, the scientific examination of animal emotion has led to little information beyond a recognition that animals have the capacity for pain and fear, and such responses as are needed for survival. Historically, prior to the rise of sciences such as ethology, interpretation of animal behaviour tended to favour a kind of minimalism known as behaviourism, in this context the refusal to ascribe to an animal a capability beyond the least demanding that would explain a behaviour; anything more than this was seen as unwarranted anthropomorphism. Put crudely, the behaviourist argument is, why should humans postulate consciousness and all its near-human implications in animals to explain some behaviour, if mere stimulus-response is a sufficient explanation to produce the same effects?

The cautious wording of Beth Dixon's 2001 paper on animal emotion exemplifies this viewpoint:

> *Recent work in the area of ethics and animals suggests that it is philosophically legitimate to ascribe emotions to non-human animals. Furthermore, it is sometimes argued that emotionality is a morally relevant psychological state shared by humans and non humans. What is missing from the philosophical literature that makes reference to emotions in non-human animals is an attempt to clarify and defend some particular account of the nature of emotion, and the role that emotions play in a characterisation of human nature. I argue in this paper that some analyses of emotion are more credible than others. Because this is so, the thesis that humans and nonhumans share emotions may well be a more difficult case to make than has been recognised thus far.*

Jeffrey Moussaieff Masson expresses a similar view:

> *While the study of emotion is a respectable field, those who work in it are usually academic psychologists who confine their studies to human emotions. The standard reference work, The Oxford Companion to Animal behaviour, advises animal behaviourists that "One is well advised to study the behaviour, rather than attempting to get at any underlying emotion."*

There is considerable uncertainty and difficulty related to the interpretation and ambiguity of emotion: an animal may make certain movements and sounds, and show certain brain and chemical signals when its body is damaged in a particular way. But does this mean an animal feels—is *aware* of—pain as we are, or does it merely mean it is programmed to act a certain way with certain stimuli? Similar questions can be asked of any activity an animal (including a human) might undertake, in principle. Many scientists regard all emotion and cognition (in humans and animals) as having a purely mechanistic basis.

Because of the philosophical questions of consciousness and mind are involved, many scientists have stayed away from examining animal and human emotion, and have studied instead, measurable brain functions, through neuroscience. The author Marc Bekoff also provided evidence of animals having emotions in his book, *The Emotional Lives of Animals*. The following is an excerpt from his book:

A few years ago my friend Rod and I were riding our bicycles around Boulder, Colorado, when we witnessed a very interesting encounter among five magpies. Magpies are corvids, a very intelligent family of birds. One magpie had obviously been hit by a car and was laying dead on the side of the road. The four other magpies were standing around him. One approached the corpse, gently pecked at it-just as an elephant noses the carcass of another elephant- and stepped back. Another magpie did the same thing. Next, one of the magpies flew off, brought back some grass, and laid it by the corpse. Another magpie did the same. Then, all four magpies stood vigil for a few seconds and one by one flew off.

Primates

Primates and in particular great apes are candidates for highly developed capabilities for empathy and theories of mind. Great apes have highly complex social systems. Young apes and their mothers have very strong bonds of attachment. Often when a baby chimpanzee or gorilla dies, the mother will carry the body around for several days. Jane Goodall has described chimpanzees as exhibiting mournful behaviour. See notably the example of the gorilla Koko, who expressed sadness over the death of her pet cat, All Ball.

Canines

Research suggests that canines can experience negative emotions in a similar manner to people, including the equivalent of certain chronic and acute psychological conditions. The classic experiment for this was Martin Seligman's foundational experiments and theory of learned helplessness at the University of Pennsylvania in 1965, as an extension of his interest in depression:

A dog that had earlier been repeatedly conditioned to associate a sound with electric shocks did not try to escape the electric shocks after the warning was presented, even though all the dog would have had to do is jump over a low divider within ten seconds, more than enough time to respond. The dog didn't even try to avoid the "aversive stimulus"; it had previously "learned" that nothing it did mattered. A follow-up experiment involved three dogs affixed in harnesses, including one that received shocks of identical intensity and duration to the others, but the lever which would otherwise have allowed the dog a degree of control was left disconnected and didn't do anything. The first two dogs quickly recovered from the experience, but the third dog suffered chronic symptoms of clinical depression as a result of this perceived helplessness.

A further series of experiments showed that (similar to humans) under conditions of long term intense psychological stress, around 1/3 of dogs do not develop learned helplessness or long term depression. Instead these animals somehow managed to find a way to handle the unpleasant situation in spite of their past experience. The corresponding characteristic in humans has been found to correlate highly with an explanatory style and optimistic attitude and lower levels of emotional rigidity regarding expectations, that views the situation as *other than* personal, pervasive, or permanent. Such studies highlighted similar distinctions between people who adapt and those who break down, under long term psychological pressure, which were conducted in the 1950s in the realm of brainwashing.

Since this time, symptoms analogous to clinical depression, neurosis and other psychological conditions have been in general accepted as being within the scope of canine emotion as well.

Psychology research has shown that human faces are asymmetrical with the gaze instinctively moving to the right side of a face upon encountering other humans to obtain information about their emotions and state. Research at the University of Lincoln (2008) shows that dogs share this instinct when meeting a human being, and only when meeting a human being (ie, not other animals or other dogs). As such they are the only non-primate species known to do so.

The existence and nature of personality traits in dogs have been studied (15329 dogs of 164 different breeds) and five consistent and stable "narrow traits" identified, described as playfulness, curiosity/fearlessness, chase-proneness, sociability and aggressiveness. A further higher order axis for shyness–boldness was also identified.

Felines

The emotions of cats have also been studied scientifically. It has been shown that cats can learn to manipulate their owners through vocalisations that are similar to the cries of human babies. Some cats learn to add a purr to the cry, which makes it less harmonious to humans and therefore harder to ignore. Individual cats learn to make these cries through operant conditioning; when a particular cry elicits a positive response from a human, the cat is more likely to use that cry in the future.

Fish

A 2007 study by the University of Guelph Scientists in Canada suggests that fish may have their own separate personalities. The study examined a group of trout that were visually identical. The study concluded that different fish within the same group exhibited different personality traits. Some fish were more willing to take risks in unknown waters than others when taken from their environment and introduced to a dark tube. Some fish were more social than others while some fish preferred being alone. Fish were also shown to have different preferences as far as eating habits.

Altruism in Animals

Altruism is a well-documented animal behaviour, which appears most obviously in kin relationships but may also be evident amongst wider social groups, in which an animal sacrifices its own well-being for the benefit of another animal.

Overview

In the science of ethology (the study of behaviour), and more generally in the study of social evolution, on occasion, some animals do behave in ways that reduce their individual fitness but increase the fitness of other individuals in the population; this is a functional definition of altruism. Research in evolutionary theory has been applied to social behaviour, including altruism. Cases of animals helping individuals to whom they are closely related can be explained by kin selection, and are not considered true altruism. Beyond the physical exertions that mothers, and in some species fathers, undertake to protect their young, extreme examples of sacrifice may occur. One example is matriphagy (the consumption of the mother by her offspring) in the spider *Stegodyphus*. Hamilton's rule describes the benefit of such altruism in terms of Wright's coefficient of relationship to the beneficiary and the benefit granted to the beneficiary minus the cost to the sacrificer. Should this sum be greater than zero a fitness gain will result from the sacrifice.

When apparent altruism is not between kin, it may be based on reciprocity. A monkey will present its back to another monkey, who will pick out parasites; after a time the roles will be reversed. Such reciprocity will pay off, in evolutionary terms, as long as the costs of helping are less than the benefits of being helped and as long as animals will not gain in the long run by "cheating" – that is to say, by receiving favours without returning them. This is elaborated on in evolutionary game theory and specifically the prisoner's dilemma as social theory.

Implications in Evolutionary Theory

Researchers on alleged altruist behaviours among animals have been ideologically opposed to the social darwinist concept of the "survival of the fittest", under the name of "survival of the nicest" — the latter being globally compatible, however, with the theory of evolution by natural selection. Insistence on such cooperative behaviours between animals was first exposed by the Russian zoologist and anarchist Peter Kropotkin in his 1902 book, *Mutual Aid: A Factor of Evolution.*

Recent developments in game theory have provided some explanations for apparent altruism, as have traditional evolutionary analyses. Among the proposed mechanisms are:

- Behavioural manipulation (for example, by certain parasites that can alter the behaviour of the host)
- Bounded rationality (for example, Herbert Simon)
- Kin selection including eusociality (see also "The Selfish Gene")
- Memes (by influencing behaviour to favourtheir own spread; see religion as a meme)
- Reciprocal altruism, mutual aid
- Sexual selection, in particular, the Handicap principle
- Reciprocity
  - — Indirect reciprocity (for example, reputation)
  - — Strong reciprocity
- Pseudo-reciprocity.

The study of altruism was the initial impetus behind George R. Price's development of the Price equation which is a mathematical equation used to study genetic evolution. An interesting example of altruism is found in the cellular slime moulds, such as *Dictyostelium mucoroides.* These protists live as individual amoebae until starved, at which point they aggregate and form a multicellular fruiting body in which some cells sacrifice themselves to promote the survival of other cells in the fruiting body. Social behaviour and altruism share many similarities to the interactions between the many parts (cells, genes) of an organism, but are distinguished by the ability of each individual to reproduce indefinitely without an absolute requirement for its neighbors. Jorge Moll and Jordan Grafman, neuroscientists at the National Institutes of Health and LABS-D'Or Hospital Network (J.M.) provided the first evidence for the neural bases of altruistic giving in normal healthy volunteers, using functional magnetic resonance imaging.

In their research, published in the Proceedings of the National Academy of Sciences USA in October, 2006, they showed that both pure monetary rewards and charitable donations activated the mesolimbic reward pathway, a primitive part of the brain that usually lights up in response to food and sex. However, when volunteers generously placed their interests of others before their own by making charitable donations, another brain circuit was selectively activated: the subgenual cortex/septal region. These structures are intimately related to social attachment and bonding in other species. Altruism, the experiment suggested, was not a superior moral faculty that suppresses basic selfish urges but rather was basic to the brain, hard-wired and pleasurable.

A new study by Samuel Bowles at the Santa Fe Institute in New Mexico, US, is seen by some as breathing new life into the model of group selection for altruism, known as "Survival of the nicest". Bowles conducted a genetic analysis of contemporary foraging groups, including Australian aboriginals, native Siberian Inuit populations and indigenous tribal groups in Africa. It was found that hunter-gatherer bands of up to 30 individuals were considerably more closely related than was previously thought. Under these conditions, thought to be similar to those of the middle and upper Paleolithic, altruism towards other group-members would improve the overall fitness of the group.

If an individual defended the group but was killed, any genes that the individual shared with the overall group would still be passed on. Early customs such as food sharing or monogamy could have levelled out the "cost" of altruistic behaviour, in the same way that income taxes redistribute income in society. He assembled genetic, climactic, archaeological, ethnographic and experimental data to examine the cost-benefit relationship of human cooperation in ancient populations. In his model, members of a group bearing genes for altruistic behaviour pay a "tax" by limiting their reproductive opportunities to benefit from sharing food and information, thereby increasing the average fitness of the group as well as their inter-relatedness. Bands of altruistic humans would then act together to gain resources from other groups at this challenging time in history.

Altruist theories in evolutionary biology were contested by Amotz Zahavi, the inventor of the signalling theory and its correlative, the handicap principle, based mainly on his observations of the Arabian Babbler, a bird commonly known for its surprising (alleged) altruistic behaviours.

Examples of Animal Altruism

- Dogs often adopt orphaned cats, squirrels, ducks and even tigers.
- Dolphins support sick or injured animals, swimming under them for hours at a time and pushing them to the surface so they can breathe.
- Wolves and wild dogs bring meat back to members of the pack not present at the kill.
- Male baboons threaten predators and cover the rear as the troop retreats.
- Gibbons and chimpanzees with food will, in response to a gesture, share their food with others of the group. Chimpanzees will help humans and conspecifics without any reward in return.
- Bonobos have been observed aiding injured or handicapped bonobos.
- Vampire bats commonly regurgitate blood to share with unlucky or sick roost mates that have been unable to find a meal, often forming a buddy system.
- Raccoons inform conspecifics about feeding grounds by droppings left on commonly shared latrines. A similar information system has been observed to be used by common ravens.
- In numerous bird species, a breeding pair receives support in raising its young from other "helper" birds, including help with the feeding of its fledglings. Some will even go as far as protecting an unrelated bird's young from predators
- Most mammal carnivores like wolves or dogs have a habit of not harming pack members below certain age, of opposite sex or in surrendering position (in case of some animals, the behaviour exists within entire species rather than one pack).
- Vervet Monkeys give alarm calls to warn fellow monkeys of the presence of predators, even though in doing so they attract attention to themselves, increasing their personal chance of being attacked.
- Walruses have been seen adopting orphans who lost their parents to predators.
- Some termites and ants release a sticky secretion by fatally rupturing a specialised gland. This autothysis altruistically aids the colony at the expense of the individual insect. For example, defending against invading ants by creating a tar baby effect.

- Meerkats often have one standing guard to warn whilst the rest feed in case of predators attack.
- African buffalo will rescue a member of the herd captured by predators.
- Lemurs of all ages and of both sexes will take care of infants unrelated to them

Cheating

Cheating is a metaphor commonly used in behavioural ecology to describe organisms that receive a benefit at the cost of others. Cheating is common in many mutualistic and altruistic relationships. Natural selection favors cheating, but there are mechanisms to regulate cheating.

Theoretical Models

The producer/scrounger game contains cheaters. Producers search for their own food while scroungers take the food others have found. Producers benefit when scroungers are common. Scroungers benefit when scroungers are rare. If there are many scroungers and few producers, the producers are easily able to find food since there is much less competition.

If there are few scroungers and many producers, the scroungers can easily cheat. The scroungers are no longer competing against each other to take the producer's food. A bird can be a scrounger. Instead of foraging for its own food, the scrounger will wait until another bird comes back with food.

The next day, the scrounger will follow the successful bird to the patch. The cheating bird reaps the benefits without putting in the effort that the producer put in to find the food.

Animal Examples

Some examples of animals that are known to cheat include: cleaner fish, social insects, ungulates, and birds. Cleaner fish have a mutualistic relationship with their clients. The cleaners eat ectoparasites off their clients. The clients benefit from being cleaned, and the cleaners benefit from acquiring food. However, some cleaners cheat by eating tissue, scales, and mucus off their clients instead of only ectoparasites. This could be detrimental to the clients and possibly turn the mutualistic relationship into commensalism or parasitism.

As for social insects, such as bees, ants, and wasps, the queen is the only individual that is supposed to reproduce. In some cases, cheating workers reproduce also. This disrupts the altruistic relationship between the queen and the workers.

Figure: *Client fish aggregate at cleaning stations, where fish such as cleaner wrasses eat the ectoparasites off the clients. The cleaner wrasse may cheat by eating tissues, scales, and mucus off the client fish instead of ectoparasites.*

Vigilance is essential for many colonial animals. The animals on the outside of the group will be vigilant, while those on the inside forage. The animals within the group also monitor each other's vigilance. A cheater may emerge in groups with vigilant animals. The cheater might try to forage the entire time, while the rest of the group switches between foraging and being vigilant.

Pain in Animals

Pain is a sensory and emotional experience often caused by intense or damaging stimuli. The International Association for the Study of Pain says pain is a conscious experience involving unpleasantness, i.e. suffering. For a creature to experience pain, by that definition, it must be capable of consciousness and suffering. The standard measure of pain in a human is that person's testimony, because only they can know the pain's quality and intensity, and the degree of suffering. Animals without human language cannot report their feelings, and whether they are conscious and capable of suffering has been a matter of some debate.

All vertebrates and some invertebrates are capable of nociception, a neural response to intense or damaging stimuli. Nociception can be observed using modern imaging techniques, and a physiological and behavioural response to nociception can be detected but, presently, there is no objective measure of suffering.

History

The idea that animals might not experience pain or suffering as humans do traces back at least to the 17th-century French philosopher, René Descartes, who argued that animals lack consciousness. Researchers remained unsure into the 1980s as to whether animals experience pain, and veterinarians trained in the U.S. before 1989 were simply taught to ignore animal pain. In his interactions with scientists and other veterinarians, Bernard Rollin was regularly asked to "prove" that animals are conscious, and to provide "scientifically acceptable" grounds for claiming that they feel pain. Some authors say that the view that animals feel pain differently is now a minority view. Academic reviews of the topic are more equivocal, noting that, although it is likely that some animals have at least simple conscious thoughts and feelings, some authors continue to question how reliably animal mental states can be determined.

In Different Species

The ability to experience pain in an animal, or another human for that matter, cannot be determined directly but it may be inferred through physiological and behavioural reactions.

Some criteria that may indicate the potential to feel pain include:

1. Has a suitable nervous system and receptors
2. Physiological changes to noxious stimuli
3. Displays protective motor reactions that might include reduced use of an affected area such as limping, rubbing, holding or autotomy
4. Has opioid receptors and shows reduced responses to noxious stimuli when given analgesics and local anaesthetics
5. Shows trade-offs between stimulus avoidance and other motivational requirements
6. Shows avoidance learning
7. High cognitive ability and sentience

Pain in Fish

The presence of pain in an animal, or another human for that matter, cannot be known for sure, but the possibility that animals experience pain can be inferred through comparative brain physiology and physical and behavioural reactions. Some specialists currently believe that all vertebrates feel pain, and that certain invertebrates, like the octopus, might too.

Animal protection advocates have raised concerns about the possible suffering of fish caused by angling. In light of recent research, some countries, like Germany, have banned specific types of fishing, and the RSPCA (which has authority in England and Wales) now formally prosecutes individuals who are cruel to fish.

Background

The idea that animals might not feel pain as human beings feel it traces back to the 17th-century French philosopher, René Descartes, who argued that animals do not experience pain and suffering, because they lack consciousness. Bernard Rollin of Colorado State University, the principal author of two U.S. federal laws regulating pain relief for animals, writes that researchers remained unsure into the 1980s as to whether animals experience pain, and veterinarians trained in the U.S. before 1989 were simply taught to ignore animal pain.

In his interactions with scientists and other veterinarians, Rollin was regularly asked to "prove" that animals are conscious, and to provide "scientifically acceptable" grounds for claiming that they feel pain. Carbone writes that the view that animals feel pain differently is now a minority one. Academic reviews of the topic are more equivocal, noting that although the argument that animals have at least simple conscious thoughts and feelings has strong support; some critics continue to question how reliably animal mental states can be determined.

Veterinary medicine uses, for actual or potential animal pain, the same analgesics and anesthetics used in humans.

Research Findings

Behaviour: Experiments by William Tavolga provide evidence that fish have pain and fear responses. For instance, in Tavolga's experiments, toadfish grunted when electrically shocked, and over time they came to grunt at the mere sight of an electrode. Additional tests conducted at both the University of Edinburgh and the Roslin Institute, in which bee venom and acetic acid was injected into the lips of rainbow trout, resulted in fish rubbing their lips along the sides and floors of their tanks, which the researchers believe was an effort to relieve themselves of pain. One researcher argues about the definition of pain used in the studies.

Figure: *Rainbow trout*

In 2003, Scottish scientists at the University of Edinburgh performing research on rainbow trout concluded that fish exhibit behaviours often associated with pain, and the brains of fish fire neurons in the same way human brains do when experiencing pain. Professor James D. Rose of the University of Wyoming criticized the study, claiming it was flawed, mainly since it did not provide proof that fish possess "conscious awareness, particularly a kind of awareness that is meaningfully like ours".

Rose argues that, since the fish brain is different from ours, fish are probably not conscious in the manner humans are, and while fish may react in a way similar to the way humans react to pain, the reactions in the case of fish have other causes. Rose had published his own opinion a year earlier arguing that fish cannot feel pain because they lack the appropriate neocortex in the brain. Studies indicating that fish can feel pain were confusing nociception (responding to threatening stimulus) with feeling pain, says Rose. "Pain is predicated on awareness. The key issue is the distinction between nociception and pain. A person who is anaesthetised in an operating theatre will still respond physically to an external stimulus, but he or she will not feel pain." However, animal behaviourist Temple Grandin argues that fish could still have consciousness without a neocortex because "different species can use different brain structures and systems to handle the same functions."

Figure 3: *Goldfish*

In a 2009 paper, Janicke Nordgreen from the Norwegian School of Veterinary Science, Joseph Garner from Purdue University, and others, published research which concluded that goldfish do feel pain, and that their reactions to pain are much like those of humans. "There has been an effort by some to argue that a fish's response to a noxious stimulus is merely a reflexive action, but that it didn't really feel pain," Garner said. "We wanted to see if fish responded to potentially painful stimuli in a reflexive way or a more clever way." The fish were divided into two groups, one given morphine and the other saline. They were then subjected to unpleasant temperatures. The fish that were given saline subsequently acted with defensive behaviours, indicating anxiety, wariness and fear, whereas those given morphine did not. Nordgreen said that the behavioural differences they found showed that fish feel both reflexive and cognitive pain. "The experiment shows that fish do not only respond to painful stimuli with reflexes, but change their behaviour also after the event," Nordgreen said. "Together with what we know from experiments carried out by other groups, this indicates that the fish consciously perceive the test situation as painful and switch to behaviours indicative of having been through an aversive experience."

Anatomy

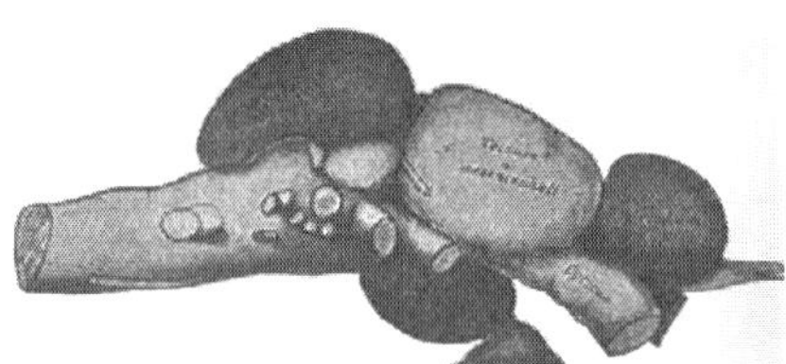

Figure: *The brain of a cod*

Nociception is the unconscious detection by the nervous system that damage is occurring somewhere. Nociceptors are sensory receptors that respond to potentially damaging stimuli by sending nerve signals to the spinal cord and brain. In 2003 Victoria Braithwaite was able to demonstrate the presence of nociceptors on the face and snout of the trout. The receptors responded to touch, heat and chemical stimulation by sending an electrical signal through the trigeminal nerve to the brain. The Norwegian Research Council is funding a three-year research project, scheduled to end in December 2011, into whether cod can feel pain. The researchers will use fMRI and EEGs to study how the cod brain works. The aim of the study is to identify the parts of the cod brain that activate when cod are exposed to potentially painful stimuli, and how those signals are processed.

Laboratory Fish

Zebrafish, native to the streams of the south eastern Himalayan region, are commonly used as a model organism in studies of vertebrate development and gene function. Zebrafish are used to study development, toxicology and toxicopathology, because the body of a young zebrafish is nearly transparent, providing unique visual access to their internal anatomy. Another extensively used model organism is the medaka, which is much sturdier than the traditional zebrafish. Medakas are easy to rear in the laboratory because of their prolific reproduction rates and short generation times. The short-lived ram cichlid is also used in laboratory studies because of its ease of breeding and predictable pattern of ageing. Sticklebacks have traditionally been used as model organism in the study of fish behaviour.

The extent to which animal research causes pain to laboratory animals is the subject of much debate. Marian Stamp Dawkins defines "suffering" in laboratory animals as the experience of one of "a wide range of extremely unpleasant subjective (mental) states." The United States Department of Agriculture defines a "painful procedure" in an animal study as one that would "reasonably be expected to cause more than slight or momentary pain or distress in a human being to which that procedure was applied."

Pain in Invertebrates

Pain in invertebrates is a problematic issue. Although there are numerous definitions of pain, almost all involve two key components. First, nociception is required. This is the ability to detect noxious stimuli which evokes a reflex response that moves the entire animal, or the affected part of its body, away from the source of the stimulus. The concept of nociception does not imply any adverse, subjective 'feeling' - it is a reflex action. The second component is the experience of 'pain' itself, or suffering, i.e. the internal, emotional interpretation of the nociceptive experience.

Figure: *A Monarch butterfly, (*Danaus plexippus*) caterpillar*

Pain is therefore a private, emotional experience. We cannot currently measure pain directly in other animals, including other humans; we can measure responses to putatively painful stimuli, but not the experience itself. To address this problem when assessing the capacity of other species to experience pain, we resort to argument-by-analogy. This is based on the principle that if an animal responds to a stimulus in a similar way to ourselves, it is likely to have had an analogous experience. Dr Chris Sherwin at the University of Bristol used this line of reasoning to question whether invertebrates have the capacity for suffering. He argued that if we stick a pin in a chimpanzee's finger and she rapidly withdraws her hand, we use argument-by-analogy and infer that like us, she felt pain. Why then, Sherwin questions, do we not infer a cockroach experiences the same when it writhes after being stuck with a pin? This argument-by-analogy approach has been revisited by Prof. Rob Elwood at the Queen's University Belfast.

The ability to experience nociception has been subject to natural selection and offers an advantage to individuals with this ability as it will reduce further harm being caused to the organism. While it might be expected therefore that nociception is widespread and robust, nociception varies across species. For example, the chemical capsaicin is commonly used as a noxious stimulus in experiments with mammals but, the African naked mole-rat, *Heterocephalus glaber*, an unusual rodent species that lacks pain-related neuropeptides (e.g., substance P) in cutaneous sensory fibres, shows a unique and remarkable lack of pain-related behaviours to acid and capsaicin. Similarly, capsaicin triggers nociceptors in some invertebrates, however, this substance is not noxious to Drosophila melanogaster. Criteria that may indicate a potential for experiencing pain include:

1. Has a suitable nervous system and receptors
2. Physiological changes to noxious stimuli
3. Displays protective motor reactions that might include reduced use of an affected area such as limping, rubbing, holding or autotomy
4. Has opioid receptors and shows reduced responses to noxious stimuli when given analgesics and local anaesthetics
5. Shows trade-offs between stimulus avoidance and other motivational requirements
6. Shows avoidance learning
7. High cognitive ability and sentience

Suitable Nervous System

Central Nervous System: One suggested reason for rejecting a pain experience in invertebrates is that invertebrate brains are too small. However, brain size does not necessarily equate to complexity of function. Moreover, weight for body-weight, the cephalopod brain is in the same size bracket as the vertebrate brain, smaller than that of birds and mammals, but as big or bigger than most fish brains.

Invertebrate nervous systems are very unlike those of vertebrates and this dissimilarity has sometimes been used to reject the possibility of a pain experience in invertebrates. In humans, the neocortex of the brain has a central role in pain and it has been argued that any species lacking this structure will therefore be incapable of feeling pain. However, it is possible that different structures may be involved in the pain experience of other animals in the way that, for example, crustacean decapods have vision despite lacking a human visual cortex.

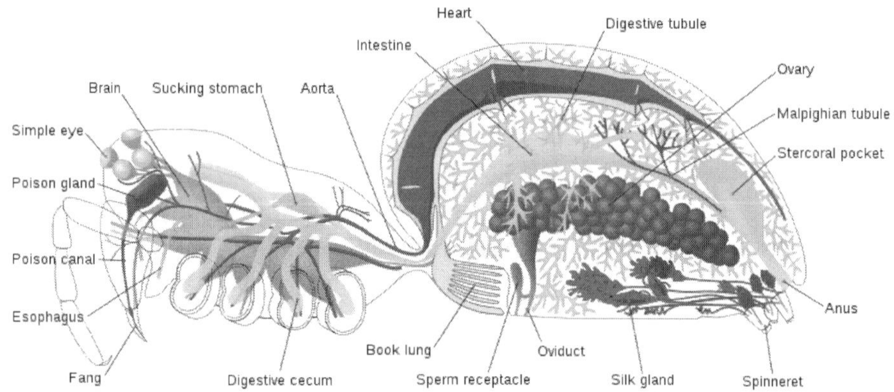

Figure: *Internal anatomy of a spider, showing the central nervous system in blue*

Two groups of invertebrates have notably complex brains: arthropods (insects, crustaceans, arachnids, and others), modern cephalopods (octopuses, squid, cuttlefish) and other molluscs. The brains of arthropods and cephalopods arise from twin parallel nerve cords that extend through the body of the animal.

Arthropods have a central brain with three divisions and large optical lobes behind each eye for visual processing. The brains of the modern cephalopods in particular are highly developed, comparable in complexity to the brains of some vertebrates (See also: Invertebrate brains). Emerging results suggest that a convergent evolutionary process has led to the selection of vertebrate-like neural organisation and activity-dependent long-term synaptic plasticity in these invertebrates. Cephalopods stand out by having a central nervous system that shares prime electrophysiological and neuroanatomical features with vertebrates like no other invertebrate taxon.

Nociceptors

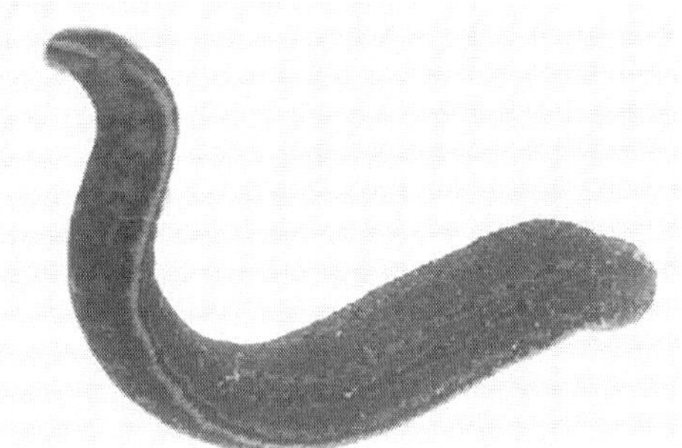

Figure 7: *Medicinal leech, Hirudo medicinalis*

Nociceptors are sensory receptors that respond to potentially damaging stimuli by sending nerve signals to the brain. Although these neurons in invertebrates may have different pathways and relationships to the central nervous system than mammalian nociceptors, nociceptive neurons in invertebrates often fire in response to similar stimuli as mammals, such as high temperature (40 C or more), low pH, capsaicin, and tissue damage.

The first invertebrate in which a nociceptive cell was identified was the medicinal leech, *Hirudo medicinalis,* which has the characteristic segmented body of an Annelida, each segment possessing a ganglion containing the T (touch), P (pressure) and N (noxious) cells. Later studies on the responses of leech neurones to mechanical, chemical and thermal stimulation motivated researchers to write *"These properties are typical of mammalian polymodal nociceptors".*

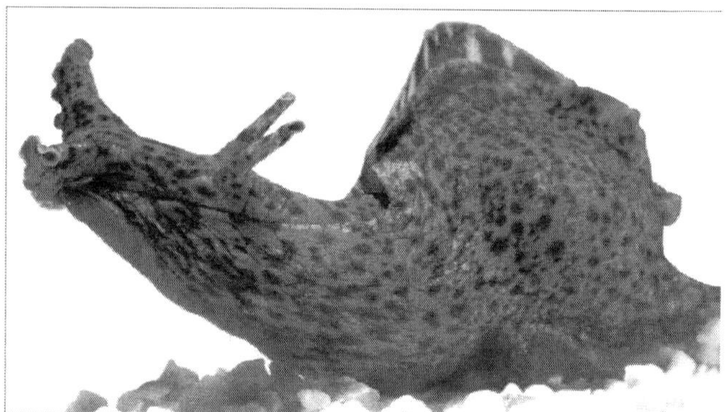

Figure: *The sea hare, Aplysia*

There have been numerous studies of learning and memory using nociceptors in the sea hare, *Aplysia.* Many of these have focused on mechanosensory neurons innervating the siphon and having their somata (bulbous end) in the abdominal ganglion (LE cells).

These LE cells display increasing discharge to increasing pressures, with maximal activation by crushing or tearing stimuli that cause tissue injury. Therefore, they satisfy accepted definitions of nociceptors. They also show similarities to vertebrate Aä nociceptors, including a property apparently unique (among primary afferents) to nociceptors — sensitization by noxious stimulation. Either pinching or pinning the siphon decreased the threshold of the LE cells firing and enhanced soma excitability.

Nociceptors have been identified in a wide range of invertebrate species, including annelids, molluscs, nematodes and arthropods.

Physiological Changes

In vertebrates, potentially painful stimuli typically produce vegetative modifications such as tachycardia, pupil dilation, defecation, changes in blood flow, respiratory patterns, arteriole blood gases, fluid and electrolyte imbalance and endocrine changes.

Figure: *The crayfish Procambarus clarkii*

At the cellular level, injury or wounding of invertebrates leads to the directed migration and accumulation of haematocytes (defence cells) and neuronal plasticity, much the same as the responses of human patients undergoing surgery or after injury. In one study, heart rate in the crayfish, *Procambarus clarkii,* decreased following claw autotomy during an aggressive encounter.

Recording physiological changes in invertebrates in response to noxious stimuli will enhance the findings of behavioural observations and such studies should be encouraged. However, careful control is required because physiological changes can occur due to noxious, but non-pain related events, e.g. cardiac and respiratory activity in crustaceans is highly sensitive and responds to changes in water level, various chemicals and activity during aggressive encounters.

Protective Motor Reactions

Invertebrates show a wide range of protective reactions to putatively painful stimuli. However, even unicellular animals will show protective responses to, for example, extremes of temperature. Many invertebrate protective reactions appear stereotyped and reflexive in action, perhaps indicating a nociceptive response rather than one of pain, but other responses are more plastic, especially when competing with other motivational systems (see section below), indicating a pain response analogous to that of vertebrates.

Mechanical Stimulation

Rather than a simple withdrawal reflex, the flatworm, *Notoplana aticola,* displays a locomotory escape behaviour following pin pricks to the posterior end. Touching the larvae of fruit flies, *D. melanogaster,* with a probe causes them to pause and move away from the stimulus, however, stronger mechanical stimulation evokes a more complex corkscrew-like rolling behaviour, i.e. the response is plastic. When a weak tactile stimulus is applied to the siphon of the sea-hare *Aplysia californica,* the animal rapidly withdraws the siphon between the parapodia. It is sometimes claimed this response is an involuntary reflex (e.g. see Aplysia gill and siphon withdrawal reflex), however, the complex learning associated with this response (see 'Learned Avoidance' below) suggests this view might be overly simplistic.

In 2001, Walters and colleagues published a report that described the escape responses of the Tobacco Hornworm caterpillar *Manduca sexta* to mechanical stimulation. These responses, particularly their plasticity, were remarkably similar to vertebrate escape responses.

> *"A set of defensive behaviour patterns in larval Manduca sexta is described and shown to undergo sensitization following noxious mechanical stimulation. The striking response is a rapid bending that accurately propels the head towards sharply poking or pinching stimuli applied to most abdominal segments. The strike is accompanied by opening of the mandibles and, sometimes, regurgitation. The strike may function to dislodge small attackers and startle larger predators. When the same*

stimuli are applied to anterior segments, the head is pulled away in a withdrawal response. Noxious stimuli to anterior or posterior segments can evoke a transient withdrawal (cocking) that precedes a strike towards the source of stimulation and may function to maximise the velocity of the strike. More intense noxious stimuli evoke faster, larger strikes and may also elicit thrashing, which consists of large, cyclic, side-to-side movements that are not directed at any target. These are sometimes also associated with low-amplitude quivering cycles. Striking and thrashing sequences elicited by obvious wounding are sometimes followed by grooming-like behaviour."

Figure: *Tobacco hornworm larva,* Manduca sexta

Tobacco Hornworm

Autotomy: Over 200 species of invertebrates are capable of using autotomy (self amputation) as an avoidance or protective behaviour including:

- land slugs *(Prophysaon)*
- sea snails *(Oxynoe panamensis)*
- crickets
- spiders
- crabs
- lobsters
- octopuses

These animals can voluntarily shed appendages when necessary for survival. Autotomy can occur in response to chemical, thermal and electrical stimulation, but is perhaps most frequently a response to mechanical stimulation during capture by a predator. Autotomy serves either to improve the chances of escape or to reduce further damage occurring to the remainder of the animal such as the spread of a chemical toxin after being stung, but the 'decision' to shed a limb or part of a body and the considerable costs incurred by this, suggests a pain response rather than simply a nociceptive reflex.

Thermal Stimulation

A heated probe (»42 °C or 108 °F) evokes a complex, corkscrew-like rolling avoidance behaviour in *Drosophila* larvae which occurs in as little as 0.4 seconds; a non-heated probe does not cause this avoidance behaviour. Land snails show an avoidance response to being placed on a hotplate (»40 °C or 104 °F) by lifting the anterior portion of the extended foot.

Chemical Stimulation

The prawn, *Palaemon elegans*, shows protective motor reactions when their antennae are treated with the irritants acetic acid or sodium hydroxide. The prawns specifically groom the treated antennae and rub them against the tank, showing they are aware of the location of the noxious stimulus on their body rather than exhibiting a generalised response to stimulation.

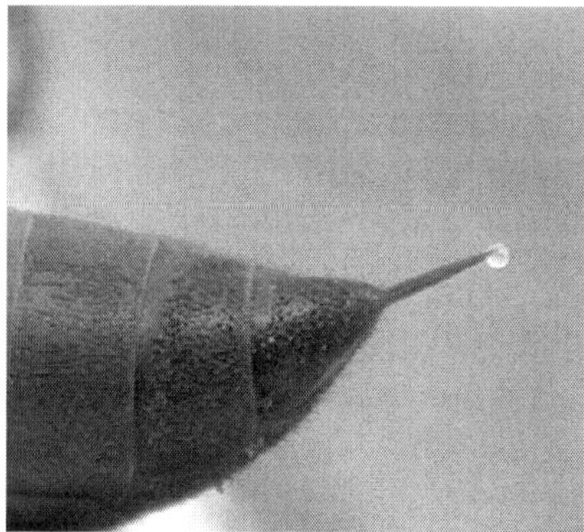

Figure: *Wasp stinger, with droplet of venom*

Under natural conditions, orb-weaving spiders (Argiope spp.) undergo autotomy (self-amputation) if they are stung in a leg by wasps or bees. Under experimental conditions, when spiders were injected in the leg with bee or wasp venom, they shed this appendage. But if they are injected with only saline, they rarely autotomize the leg, indicating it is not the physical insult or the ingress of fluid *per se* that causes autotomy. Even more interestingly, spiders injected with venom components which cause injected humans to report pain (serotonin, histamine, phospholipase A2 and melittin) autotomize the leg, but if the injections contain venom components which do not cause pain to humans, autotomy does not occur.

Electrical Stimulation

The sea-slug, *Tritonia diomedia,* possesses a group of sensory cells, "S-cells", situated in the pleural ganglia, which initiate escape swimming if stimulated by electric shock. Similarly, Mantis shrimps, *Squilla mantis,* show avoidance of electric shocks with a strong tail-flick escape response. Both these responses appear to be rather fixed and reflexive, however, other studies indicate a range of invertebrates exhibit considerably more plastic responses to electric shocks.

Because of their soft bodies, Hermit crabs rely on shells for their survival, but, when they are given small electric shocks within their shells, they evacuate these. The response, however, is influenced by the attractivness of the shell; more preferred shells are only evacuated when the crabs are given a higher voltage shock, indicating this is not a simple, reflex behaviour.

In studies on learning and the Aplysia gill and siphon withdrawal reflex, *Aplysia* received an electric shock on the siphon each time their gill relaxed below a criterion level. The *Aplysia* learned to keep their gills contracted above the criterion level - an unlikely outcome if the response was due to a nociceptive experience.

Drosophila feature widely in studies of invertebrate nociception and pain. It has been known since 1974 that these fruit-flies can be trained with sequential presentations of an odour and electric shock (odour-shock training) and will subsequently avoid the odour because it predicts something 'bad'. A similar response has been found in the larvae of this species. In an intruiging study, *Drosophila* learned two kinds of prediction regarding a 'traumatic' experience. If an odour preceded an electric shock during training, it predicted shock and the flies subsequently avoided it. When the sequence of events during training was reversed, i.e. odour followed shock, the odour predicted relief from shock and flies approached it. The authors termed this latter effect 'relief' learning.

Many invertebrate species learn to withdraw from, or alter their behaviour in response to, a conditioned stimulus when this has been previously paired with an electric shock - cited by Sherwin - and include snails, leeches, locusts, bees and various marine molluscs.

If vertebrate species are used in studies on protective or motor behaviour and they respond in similar ways to those described above, it is usually assumed that the learning process is based on the animal experiencing a sensation of pain or discomfort from the stimulus, e.g.

an electric shock. Argument-by-analogy suggests an analogous experience occurs in invertebrates.

Opioid Receptors, Effects of Local Anaesthetics or Analgesics

In vertebrates, opiates modulate nociception and opioid receptor antagonists, e.g. naloxone and CTOP, reverse this effect. So, if opiates have similar effects in invertebrates as vertebrates, they should delay or reduce any protective response and the opioid antagonist should counteract this. It has been found that molluscs and insects have opioid binding sites or opioid general sensitivity. Certainly there are many examples of neuropeptides involved in vertebrate pain responses being found in invertebrates, for example, endorphins have been found in platyhelminthes, molluscs, annelids, crustaceans and insects (see). It should be noted, however, that apart from analgesia, there are other effects of exogenous opiates specifically being involved in feeding behaviour and activation of immunocytes. These latter functions might explain the presence of opioids and opioid receptors in extremely simple invertebrates and unicellular animals.

Nematodes

Nematodes avoid extremes of temperature. Morphine increases the latency of this defensive response in the parasitic *Ascaris suum*. In a study on the effects of opiates in *Caenorhabditis elegans,* 76% of a non-treated group exhibited a rapid, reflexive withdrawl to heat, whereas 47%, 36% and 39% of morphine, endomorphin 1 and endomorphin 2 treated worms (respectively) withdrew. These effects were reversed with the opioid receptor antagonists naloxone and CTOP, leading the authors to conclude that thermonocifensive behaviour in *C. elegans* was modulated by opioids.

Molluscs

Figure: *Helix pomatia, a species of land snail*

Slugs and snails posess an opioid receptor system. In experiments on different terrestrial snails, morphine prolonged the latency of the snails' raising their foot in response to being placed on a hot (40°C) surface. The analgesic effects of the morphine were eliminated by naloxone as is seen in humans and other vertebrates. There was also habituation to morphine. Snails administered with morphine for four days did not differ from the control ones in tests on pain sensitivity and analgesia was achieved only at a higher dose.

Crustaceans

A selection of crustaceans that show responses to analgesics and their agonists:

Figure: *The Mud-flat crab,* Chasmagnathus

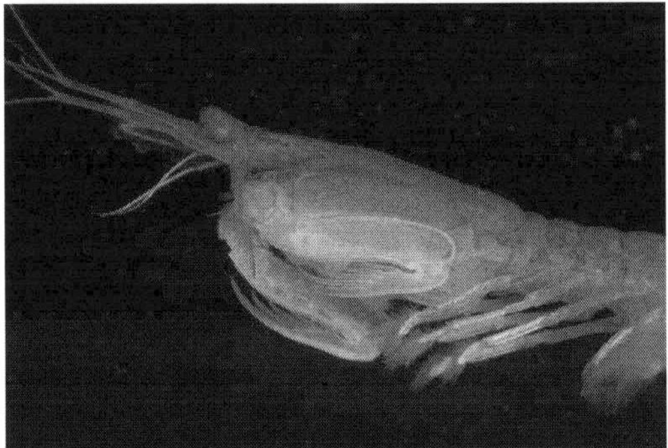

Figure: *The Mantis shrimp, Squilla mantis*

Evidence of the capacity for invertebrates to experience nociception and pain has been widely studied in crustaceans. In the crab *Chasmagnathus granulatus,* electric shocks delivered via small holes in the carapace elicited a defensive threat display. Injection of morphine reduced the crabs' sensitivity to the shock in a dose-dependent manner, with the effect declining with increasing duration between morphine injection and shock. Naloxone injection inhibited the effects of morphine, as is seen in vertebrates. Morphine also had inhibitory effects on the escape tail-flick response to electric shock in the mantis shrimp,

Squilla mantis, that was reversed by naloxone, indicating that the effect is found in crustacean groups other than decapods. When the irritants acetic acid or sodium hydroxide were applied to the antennae of Grass prawns, *Penaeus monodon*, there was an increase in rubbing and grooming of the treated areas which was not seen if they had previously been treated with a local anaesthetic, benzocaine, however, the benzocaine did not eliminate the level of rubbing seen in response to mechanical stimulation with forceps. There was no effect of benzocaine on the general locomotion of the prawns, so the reduction in rubbing and grooming was not simply due to inactivity of the animal. Another local anaesthetic, xylocaine, reduced the stress of eyestalk ablation in female Whiteleg shrimps, *Litopenaeus vannamei*, as indicated by levels of feeding and swimming.

It has not always been possible to replicate these findings in crustaceans. In one study, three decapod crustacean species, Louisiana red swamp crayfish, white shrimp and grass shrimp were tested for nociceptive behaviour by applying sodium hydroxide, hydrochloric acid, or benzocaine to the antennae. This caused no change in behaviour in these three species compared to controls. Animals did not groom the treated antenna, and there was no difference in movement of treated individuals and controls. Extracellular recordings of antennal nerves in the Louisiana red swamp crayfish revealed continual spontaneous activity, but no neurons that were reliably excited by the application of sodium hydroxide or hydrochloric acid. The authors concluded there was no behavioural or physiological evidence that the antennae contained specialised nociceptors that responded to pH. It could be argued that differences in the findings between studies may be due to responses to extreme pH being inconsistently evoked across species.

It has been argued that the analgesic effects of morphine should not be used as a criterion of the ability of animals, at least crustaceans, to experience pain. In one study, Shore crabs, *Carcinus maenas* received electric shocks in a preferred dark shelter but not if they remained in an unpreferred light area. Analgesia from morphine should have enhanced movement to the preferred dark area because the crabs would not have experienced 'pain' from the electric shock. However, morphine inhibited rather than enhanced this movement, even when no shock was given. Morphine produced a general effect of non-responsiveness rather than a specific analgesic effect, which could also explain previous studies claiming analgesia. However, the researchers argued that other systems such as the enkephalin or steroid systems might be used in pain modulation by crustaceans and that behavioural responses should be considered rather than specific physiological and morphological features.

Insects

Figure: *The house cricket, Acheta domestica*

Morphine extends the period that crickets avoided the heated surface of a hotplate.

Trade-offs between Stimulus Avoidance and other Motivational Requirements

This is a particularly important criterion for assessing whether an animal has the capacity to experience pain rather than only nociception. Nociceptive responses do not require consciousness or higher neural processing; this results in relatively fixed, reflexive actions. However, the experience of pain does involve higher neural centres which also take into account other factors of relevance to the animal, i.e. competing motivations. This means that a response to the experience of pain is likely to be more plastic than a nociceptive response when there are competing factors for the animal to consider.

Robert Elwood and Mirjam Appel at the Queen's University of Belfast argue that pain may be inferred when the responses to a noxious stimulus are not reflexive but are traded off against other motivational requirements, the experience is remembered and the situation is avoided in the future. They investigated this by giving hermit crabs small electric shocks within their shells. Only crabs given shocks evacuated their shells indicating the aversive nature of the stimulus, but fewer crabs evacuated from a preferred species of shell demonstrating a motivational trade-off. Most crabs, however, did not evacuate at the shock level used, but when these shocked crabs were subsequently offered a new shell, they were more likely to approach and enter the new shell. They approached the new shell more quickly, investigated it for a shorter time and used fewer cheliped probes within the aperture prior to moving in. This demonstrates the experience of the electric shock altered future behaviour in a manner consistent with a marked shift in motivation to get a new shell to replace the one previously occupied.

Learned Avoidance

Learning to avoid a noxious stimulus indicates that prior experience of the stimulus is remembered by the animal and appropriate action taken in the future to avoid or rduce potential damage. This type of response is therefore not the fixed, reflexive action of nociceptive avoidance.

Habituation and Sensitization

Habituation and sensitisation are two simple, but widespread, forms of learning. Habituation refers to a type of non-associative learning in which repeated exposure to a stimulus leads to decreased responding. Sensitization is another form of learning in which the progressive amplification of a response follows repeated administrations of a stimulus.

When a tactile stimulus is applied to the skin of *Aplysia californica,* the animal withdraws the siphon and gill between the parapodia. This defensive withdrawal, known as the Aplysia gill and siphon withdrawal reflex, has been the subject of much study on learning behaviour. Generally, these studies have involved only weak, tactile stimulation and are therefore more relevant to the question of whether invertebrates can experience nociception, however, some studies have used electric shocks to examine this response (See sections on "Electrical stimulation" and "Operant conditioning").

Other researchers working with Aplysia were sufficiently impressed about the similarity between invertebrate and mammalian responses to write - *"Persistent nociceptive sensitization of nociceptors in* Aplysia *displays many functional similarities to alterations in mammalian nociceptors associated with the clinical problem of chronic pain. Moreover, in Aplysia and mammals the same cell signaling pathways trigger persistent enhancement of excitability and synaptic transmission following noxious stimulation, and these highly conserved pathways are also used to induce memory traces in neural circuits of diverse species"*

Location Avoidance

Avoidance learning was examined in the crab, *Chasmagnathus granulatus*, by placing the animals in a the dark compartment of a double-chamber device and allowing them to move towards a light compartment. Experimental crabs received a shock in the light compartment, whilst controls did not. After 1 min, both experimental and control crabs were free to return to the dark compartment. The learned outcome was not a faster escape response to the stimulus but rather refraining from re-entering the light compartment. A single trial was enough to establish an association between light and shock that was detected up to 3 hours later.

Studies on crayfish, *Procambarus clarkia*, demonstrated that they learned to associate the turning on of a light with a shock that was given 10 seconds later. They learned to respond by walking to a safe area in which the shock was not delivered. However, this only occurred if the crayfish were facing the area to which they could retreat to avoid the shock. If they were facing away from the safe area the animal did not walk but responded to the shock by a tail-flick escape response. Despite repeated pairings of light and shock the animals did not learn to avoid the shock by tail-flicking in response to light. Curiously, when the animals that had experienced shocks whilst facing away from the safe area were subsequently tested facing towards the safe area they showed a very rapid avoidance of the shock upon the onset of the light. Thus, they seemed to have learned the association although they had not previously used it to avoid the shock - much like mammalian latent learning. These studies show an ability in decapods that fulfils several criteria for pain experience rather than nociception.

Conditioned Suppression

Figure: *A drone bee*

Honeybees extend their proboscis when learning about novel odours. In one study on this response, bees learnt to discriminate between two odours, but then learned to suppress the proboscis extension response when one of the odours was paired with an electric shock. This indicates the sensation was aversive to the bee, however, the response was plastic rather than simply reflexive, indicating pain rather than nociception.

Operant Conditioning

Operant studies using vertebrates have been conducted for many years. In such studies, an animal operates or changes some part of the environment to gain a positive reinforcement or avoid a negative one. In this way, animals learn from the consequence of their own actions, i.e. they use an internal predictor. Operant responses indicate a voluntary act; the animal exerts control over the frequency or intensity of its responses, making these distinct from reflexes and complex fixed action patterns. A number of studies have revealed surprising similarities between vertebrates and invertebrates in their capacity to use operant responses to gain positive reinforcements, but also to avoid negative reinforcement that in vertebrates we would describe as 'pain'.

Figure: *Underside of a snail climbing a blade of grass, showing the muscular foot*

Snail

It has been shown that snails will operate a manipulandum to electrically self-stimulate areas of their brain. Balaban and Maksimova surgically implanted fine wire electrodes in two regions of the brains of snails (Helix sp.). To receive electrical stimulation of the brain, the snail was required to displace the end of a rod. When pressing the rod delivered self-stimulation to the mesocerebrum (which is involved in sexual activity) the snails increased the frequency of operating the manipulandum compared to the baseline spontaneous frequency of operation. However, when stimulation was delivered to the parietal ganglion, the snails decreased the frequency of touching the rod compared to the baseline spontaneous frequency. These increases and decreases in pressing are positive and negative reinforcement responses typical of those seen with vertebrates.

Aplysia

To examine the gill and siphon withdrawl response to a putatively painful stmulus, *Aplysia* were tested in pairs. During the initial training period, the experimental animal received a siphon shock each time its gill relaxed below a criterion level, and the yoked control animal received a shock whenever the experimental animal did, regardless of its own gill position. The experimental animals spent more time with their gills contracted above the criterion level than did the control animals during each period, demonstrating operant conditioning.

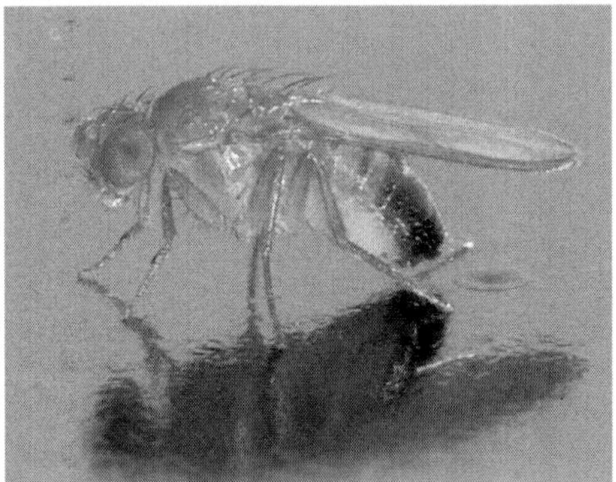

Figure: *Drosophila melanogaster*

Drosophila

A fly-controlled heat-box has been designed to study operant conditioning in several studies of *Drosophila*. Each time a fly walks into the designated half of the tiny dark chamber, the whole space is heated. As soon as the animal leaves the punished half, the chamber temperature reverts to normal. After a few minutes, the animals restrict their movements to one-half of the chamber, even if the heat is switched off.

A *Drosophila* flight simulator has been used to examine operant conditioning. The flies are tethered in an apparatus that measures the yaw torque of their flight attempts and stabilises movements of the panorama. The apparatus controls the fly's orientation based on these attempts. When the apparatus was set up to direct a heat beam on the fly if it "flew" to certain areas of its panorama, the flies learned to prefer and avoid certain flight orientations in relation to the surrounding panorama. The flies "avoided" areas that caused them to receive heat. These experiments show that *Drosophila* can use operant behaviour and learn to avoid noxious stimuli. However, these responses were plastic, complex behaviours rather than simple reflex actions, consistent more with the experience of pain rather than simply nociception.

Cognitive Abilities

It could be argued that a high cognitive ability is not necessary for the experience of pain, otherwise, it could be argued that humans with less cognitive capacity have a lower likelihood of experiencing pain. However, most definitions of pain indicate some degree of cognitvie ability. Several of the learned and operant behaviours described above indicate that invertebrates have high cognitive abilities. Other examples include -

- Social transmission of information during the waggle dance of honeybees.
- Idiothetic orientation by spiders, i.e. they memorize information about their previous movements.
- Detour behaviour in which spiders choose to take an indirect route to a goal rather than the most direct route, thereby indicating flexibility in behaviour and route planning, and possibly insight learning.
- Conceptualisation in the honeybee, *Apis mellifera.*
- Problem solving in leafcutter ants, *Atta colombica.*
- Numeracy in the yellow mealworm beetle, *Tenebrio molitor,* and honeybee.

In Medicine and Research

Veterinary Medicine: Veterinary medicine uses, for actual or potential animal pain, the same analgesics and anesthetics as used in humans.

Dolorimetry

Dolorimetry (*dolor*: Latin: pain, grief) is the measurement of the pain response in animals, including humans. It is practiced occasionally in medicine, as a diagnostic tool, and is regularly used in research into the basic science of pain, and in testing the efficacy of analgesics. Non-human animal pain measurement techniques include the paw pressure test, tail flick test and hot plate test.

Laboratory Animals

Animals are kept in laboratories for a wide range of reasons, some of which may involve pain, suffering or distress, whilst others (e.g. many of those involved in breeding) will not. The extent to which animal testing causes pain and suffering in laboratory animals is the subject of much debate. Marian Stamp Dawkins defines "suffering" in laboratory animals as the experience of one of "a wide range of extremely unpleasant subjective (mental) states." The United States Department of Agriculture defines a "painful procedure" in an animal study as one that would "reasonably be expected to cause more than slight or momentary pain or distress in a human being to which that procedure was applied." Some critics argue that, paradoxically, researchers raised in the era of increased awareness of animal welfare may be inclined to deny that animals are in pain simply because they do not want to see themselves as people who inflict it. Animal research with the potential to cause pain is regulated by the Animal Welfare Act of 1966 in the US, and research likely to cause "pain, suffering, distress or lasting harm" is regulated by the Animals (Scientific Procedures) Act 1986 in the UK.

Severity Scales

To date (2011), eleven countries have national severity classification systems relating to pain and suffering experienced by animals used in research: Australia, Canada, Finland, Germany, The Republic of Ireland, The Netherlands, New Zealand, Poland, Sweden, Switzerland, and the UK. The US also has a mandated national scientific animal-use classification system, but it is markedly different from other countries in that it reports on whether pain-relieving drugs were required and/or used. The first severity scales were implemented in 1986 by Finland and the UK. The number of of severity catagories ranges between 3 (Sweden and Finland) and 9 (Australia). In the UK, research projects are classified as "mild", "moderate", and "substantial" in terms of the suffering the researchers conducting the study say they may cause; a fourth category of "unclassified" means the animal was anesthetised and killed without recovering consciousness. It should be remembered that in the UK system, many research projects (e.g. transgenic breeding, feeding distasteful food) will require a license under the Animals (Scientific Procedures) Act 1986, but may cause little or no pain or suffering. In December 2001, 39 percent (1,296) of project licenses in force were classified as "mild", 55 percent (1,811) as "moderate", two percent (63) as "substantial", and 4 percent (139) as "unclassified". In 2009, of the project licenses issued, 35 percent (187) were classified as "mild", 61 percent (330) as "moderate", 2 percent (13) as "severe" and 2 percent (11) as unclassified.

In the US, the *Guide for the Care and Use of Laboratory Animals* defines the parameters for animal testing regulations. It states, "The ability to experience and respond to pain is widespread in the animal kingdom...Pain is a stressor and, if not relieved, can lead to unacceptable levels of stress and distress in animals." The *Guide* states that the ability to recognise the symptoms of pain in different species is essential for the people caring for and using animals. Accordingly, all issues of animal pain and distress, and their potential treatment with analgesia and anesthesia, are required regulatory issues for animal protocol approval.

Pain in Crustaceans

The question of whether or not crustaceans can experience pain is unresolved. Because of the ambiguous nature of pain, most people who contend that crustaceans do have this capacity approach the issue using 'argument by analogy' – that is, they hold that certain similarities between crustacean and human biology or behaviour warrant an assumption that crustaceans can feel pain.

In vertebrates, endogenous opioids are neurochemicals that moderate pain by interacting with opioid receptors. Opioid peptides and opioid receptors occur naturally in crustaceans, and although "at present no certain conclusion can be drawn", some have interpreted their presence as an indication that crustaceans may be able to experience pain. Lobsters' opioids may "mediate pain in the same way" as in vertebrates.

Background

In February 2005, a review of the literature by the Norwegian Scientific Committee for Food Safety tentatively concluded that "it is unlikely that [lobsters] can feel pain," though they note that "there is apparently a paucity of exact knowledge on sentience in crustaceans, and more research is needed." This conclusion is based on the lobster's simple nervous system. The report assumes that the violent reaction of lobsters to boiling water is a reflex to noxious stimuli.

However, review by the Scottish animal rights group Advocate for Animals released in the same year reported that "scientific evidence ... strongly suggests that there is a potential for [lobsters] to experience pain and suffering," primarily because lobsters (and other decapod crustaceans) "have opioid receptors and respond to opioids (analgesics such as morphine) in a similar way to vertebrates," indicating that lobsters' reaction to injury changes when painkillers are applied. The similarities in lobsters' and vertebrates' stress systems and behavioural responses to noxious stimuli were given as additional evidence for their capacity for pain.

A 2007 study at Queen's University, Belfast, suggested that crustaceans do feel pain. In the experiment, when the antennae of prawns were rubbed with sodium hydroxide or acetic acid, the animals showed increased grooming of the afflicted area and rubbed it more against the side of the tank. Moreover, this reaction was inhibited by a local anesthetic, even though control prawns treated with only anesthetic did not show reduced activity. Robert Elwood, who headed the study, argues that sensing pain is crucial to prawn survival, because it encourages them to avoid damaging behaviours. Some scientists responded, saying the rubbing may reflect an attempt to clean the affected area.

In a subsequent 2009 study, Elwood and Mirjam Appel showed that hermit crabs make motivational tradeoffs between electric shocks and the quality of the shells they inhabit. In particular, as hermit crabs are shocked more intensely, they become increasingly willing to leave their current shells for new shells, and they spend less time deciding whether to enter those new shells. Moreover, because the researchers did not offer the new shells until after the electrical stimulation had ended, the change in motivational behaviour was the result of memory of the noxious event, not an immediate reflex.

Morphine, an analgesic, and naloxone, an opioid receptor antagonist, may affect the estuarine crab *Neohelice granulata* in much the same way they affect vertebrates: injections of morphine into crabs produced a dose-dependent reduction of their defensive response to an electric shock. (However, the attenuated defensive response could originate from either the analgesic or sedative properties of morphine, or both.) These findings have been replicated for other invertebrate species, but similar data is not yet available for lobsters.

Chapter 8

Animal Communication

Animal communication is any behaviour on the part of one animal that has an effect on the current or future behaviour of another animal. The study of animal communication, is sometimes called Zoosemiotics (defined as the study of sign communication or semiosis in animals; distinguishable from anthroposemiotics, the study of human communication) has played an important part in the methodology of ethology, sociobiology, and the study of animal cognition.

Animal communication, and the understanding of the animal world in general, is a rapidly growing field. Even in the 21st century, many prior understandings related to diverse fields such as personal symbolic name use, animal emotions, animal culture and learning, and even sexual conduct, long thought to be well understood, have been revolutionised.

Forms of Communication

- *Gestures:* The best known form of communication involves the display of distinctive body parts, or distinctive bodily movements; often these occur in combination, so a distinctive movement acts to reveal or emphasize a distinctive body part. For example, the presentation of a parent Herring Gull's bill to its chick signals feeding time. Like many gulls, the Herring Gull has a brightly coloured bill, yellow with a red spot on the lower mandible near the tip. When it returns to the nest with food, the parent stands over its chick and taps the bill on the ground in front of it; this elicits a begging response from a hungry chick (pecking at the red spot), which stimulates the parent to regurgitate food in front of it. The complete signal therefore involves a distinctive morphological feature (body part), the red-spotted bill, and a distinctive movement (tapping towards the ground) which makes the red spot highly visible to the chick. Congruently, some cephalopods, such as the octopus, have specialised skin cells that can change the apparent colour, opacity, and reflectiveness of their skin. In addition to being used for camouflage, rapid changes in skin colour are used while hunting and in courtship rituals. While all primates use some form of gesture, Frans de Waal came to the conclusion that apes and humans are unique in that only they are able use intentional gestures to communicate. He tested the hypothesis of gesture evolving into language by studying the gestures of bonobos and chimps.

- *Facial Expression:* Facial gestures play an important role in animal communication. Dogs for example express anger through a snarling and showing their teeth. In alarm their ears will perk up. When fearful a dog will pull back their ears, expose teeth slightly and squint eyes. Jeffrey Mogil studied the facial expressions of mice during increments of increasing pain. What they found were five recognisable facial expressions; orbital tightening, nose and check bulge, and changes in ear and whisker carriage.
- *Gaze Following:* Coordination among social animals is facilitated by monitoring of each others' head and eye orientation. Long recognised in human developmental studies as an important component of communication, there has recently begun to be much more attention on the abilities of animals to follow the gaze of those they interact with, whether members of their own species or humans. Studies have been conducted on apes, monkeys, dogs, birds, and tortoises, and have focused on two different tasks: "following another's gaze into distant space" and "follow[ing] another's gaze geometrically around a visual barrier e.g. by repositioning themselves to follow a gaze cue when faced with a barrier blocking their view". The first ability has been found among a broad range of animals, while the second has been demonstrated only for apes, dogs (and wolves), and corvids (ravens), and attempts to demonstrate this "geometric gaze following" in marmosets and ibis gave negative results. Researchers do not yet have a clear picture of the cognitive basis of gaze following abilities, but developmental evidence indicates that "simple" gaze following and "geometric" gaze following are likely to rely on distinct cognitive foundations.

- Vocalisation: Many animals communicate through vocalisations. Communication through vocalisation is essential for many tasks including mating rituals, warning calls, conveying location of food sources, and social learning. Male mating calls are used to signal the female and to beat competitors in species such as hammer-headed bats, red deers, humpback whales and elephant seals . In whale species Whale song has been found to have different dialects based on location. Other instances of communication include the warning cries of the Campbell monkey, the territorial calls of gibbons, the use of frequency in Greater Spear-nosed bats to distinguish between groups .
- Olfactory communication: Less obvious (except in a few cases) is olfactory communication. Many mammals, in particular, have glands that generate distinctive and long-lasting smells, and have corresponding behaviours that leave these smells in places where they have been. Often the scented substance is introduced into urine or feces. Sometimes it is distributed through sweat, though this does not leave a semi-permanent mark as scents deposited on the ground do. Some animals have glands on their bodies whose sole function appears to be to deposit scent marks: for example Mongolian gerbils have a scent gland on their stomachs, and a characteristic ventral rubbing action that deposits scent from it. Golden hamsters and cats have scent glands on their flanks, and deposit scent by rubbing their sides against objects; cats also have scent glands on their foreheads. Bees carry with them a pouch of material from the hive which they release as they reenter, the smell of which indicates that they are a part of the hive and grants their safe entry. Ants use pheromones to create scent trails to food as well as for alarm calls, mate attraction and to distinguish between colonies. Additionally, they have pheromones that are used to confuse an enemy and manipulate them into fighting with themselves .
- Bioluminescence, common in the oceans deeps and with fireflies.
- Electro communication: A rarer form of animal communication is electrocommunication. It is seen primarily in aquatic life, though some mammals, notably the platypus and echidnas are capable of electroreception and thus theoretically of electrocommunication.

Agonistic Behaviour

In ethology, agonistic behaviour is any social behaviour related to fighting. Thus it is broader than aggressive behaviour because it includes not only actual aggression but also threats, displays, retreats, placating aggressors, and conciliation. The term was coined by Scott and Fredericson in 1951. Agonistic behaviour is seen in many animal species because resources including food, shelter, and mates are often limited.

Some forms of agonistic behaviour are between contestants who are competing for access to the same resources, such as food or mates. Other times it involves tests of strength or threat display that make animals look large and more physically fit, a display that may allow it to gain the resource before an actual battle takes place. Although agonistic behaviour varies among species, agonistic interaction consists of three kinds of behaviours: threat, aggression, and submission. These three behaviours are functionally and physiologically interrelated with aggressive behaviour yet fall outside the narrow definition of aggressive behaviour. While any one of these divisions of behaviours may be seen alone in an interaction between two animals, they normally occur in sequence from start to end. Depending on the availability and importance of a resource, behaviours can range from a fight to the death or a much safer ritualistic behaviour, though ritualistic or display behaviours are the most common form of agonistic behaviours.

Prediction of Winning

The type of agonistic behaviour observed, whether it be aggressive or submissive, all depend on the likelihood of winning. For instance, size is usually a good predictor of fighting success, and many animals will display to flaunt their size. Animals are better able to assess their next form of agonistic action by judging the opponent's size and if they are likely to win a fight if a physical fight were to occur.

Agonistic Behaviour in Stalk-eyed Flies

In aggressive behaviour by male stalk-eyed flies (Diopsidae), the males "square off" by displaying their eyes. Females show a strong preference for mating with males with longer eyestalks. Due to the female preference, males have evolved to compete with each other for mating rights. In the threat display the two flies face each other head-to-head, with their forelegs spread outward and parallel to the eyestalks. This behaviour allows each individual to judge the distance between its competitor's eyes. Eyestalk length increases with body size, and males with shorter eyestalks usually will retreat. A further distance between the eyes conveys a bigger body size, and a better chance of winning.

Avoidance

Physical fighting is actually rare between animals. It would seem that normally the more aggressive an animal is, the more it has to gain. However, in a normal scenario if an animal is too aggressive it might face an unacceptably high cost such as severe injury or death. Unless an animal has a sure indication that they will win without injury, or the resources are valuable enough for the risk of death, animals usually avoid fighting. An animal must weigh the relative costs and benefits of fighting. If the costs are too high, avoiding a fight is preferable.

Ritual Display

For animals, display is any behaviour modified by evolution that is used to convey information. Animals display particular signs, which recipients can use to infer something about the mental and physical state of the first animal. To avoid the heavy cost of fighting, animals have evolved sophisticated rituals, which they use to bluff their opponents into backing down or fleeing. The cost-benefit model of display makes three assumptions: (1) type of display varies depending on the cost; (2) the risk of the display increases as the effectiveness of display increases: and (3) the value of resource being disputed over determines the choice of display used. Animals have evolved to use their physical attributes as a display of ability. If contests can be resolved with ritual display, fighting is not needed. Display can be used to dispute for mates, territory, and food through symbolic gestures instead of battles to the death. If an animal can display without fighting that he is more physically fit than his opponent, he will have gained more than he would have if he had fought and in the process possibly been injured.

*Male Grey Catbird (*Dumetella carolinensis*)*

Male grey catbirds fluff their feathers and spread and lower their tails to defend their territory when threatened by another male. The bird that is capable of puffing up and appearing to be the biggest will win the territory.

*Western Gorilla (*Gorilla gorilla*)*

Male western gorillas display a wide range of both vocal and gestural communications when threatened by an opponent. A silverback (alpha male) will start hooting, throwing, chest pounding, leg kicks, and sideways running when approached by another male. This is done to intimidate the opponent and show physical abilities without actually making any physical contact.

Threats

Threat behaviour is any behaviour that signifies hostility or intent to attack another animal. Threat behaviour is meant to cause the opponent to back down and leave. While ritual display can be used for an array of reasons or communicative purposes, threat distinctly is meant for hostility and is the last step before fighting or submission. Threat does not involve physical contact with another animal. Any threat behaviour most often elicits other agonistic behaviour in the recipient.

This initiation of threat will result in a display of physical attributes, a fight, or submission; the behaviour or sequence of behaviours depends on what resources are being fought over and each individual's chance of winning against his opponent. In any animal species, threat always contains components of attack and fleeing, which expresses an animal's readiness and likelihood of winning. An intimidation display with a means to threat are exhibited through: hair bristling, feather ruffling, raising skin folds and crest, teeth displaying, horn displaying, making sound, etcetera.

Chlamydosaurus Kingii

Chlamydosaurus kingii, an Australian agamid lizard, uses its frill as a way to display size and aggression to opponents. It is one of the largest and most notable displays seen in the animal kingdom. In comparison to its body size, the frill can flare out to make the lizards head look several times bigger, and it displays bright orange and red scales. Males of *C. kinggi* fight and display frills often during the mating seasons. The male ritualistic display includes repeated partial erections of the frill, head bobbing, tail lashing, and waving of forelimbs.

Agonistic Fighting

Actual fighting in contests is rare because of the risk of injury. It is most likely to occur when individuals are similarly sized, or when reproduction and survival without the contested resource is less likely. Even when agonism escalates to actual fighting, restraint may be used. This is the case in fights among some male venomous snakes for females. They wrestle, but refrain from biting.

*Agonistic Fighting in Black Mambas (*Dendroaspis polylepius*)*

Agnostic fighting for black mambas involves a wrestling match in which opponents attempt to pin each other's head repeatedly to the ground. Fights normally last a few minutes but can extend to over an hour or more. The purpose of fighting is to secure mating rights to receptive females nearby during the breeding season.

Submissive Behaviour

Submissive behaviour involves an individual indicating by an act or posture that it will not challenge a dominant individual in a social group. Submissive behaviours are part of the maintenance of a dominance hierarchy of cooperating individuals in a social group that have overlapping but not entirely coincident interests.

*Submission in Bearded Dragons (*Pogona vitticeps*)*

Communication between animals is often achieved by adding a succession of behavioural responses to a display. Social interactions among bearded dragons (*Pogona vitticeps*) consist of a unique set of movements that convey visual signals and communication. Waving is one of the most visible signs of submission one lizard can display to another Lizard. The lizard will rest on three of its legs and raise one of the front arms and then slowly wave the arm in a circular motion. This circular motion, along with the dragon puffing up slightly, shows submission. This display is seen between opponents, as well as from adolescents towards adults.

Territory

In ethology the term territory refers to any sociographical area that an animal of a particular species consistently defends against conspecifics (and, occasionally, animals of other species). Animals that defend territories in this way are referred to as territorial.

Classic Territories

Territorial animals defend areas that contain a nest, den or mating site and sufficient food resources for themselves and their young. Defense rarely takes the form of overt fights: more usually there is a highly noticeable display, which may be visual (as in the red breast of the robin), auditory (as in much bird song, or the calls of gibbons) or olfactory, through the deposit of scent marks. Many territorial mammals use scent-marking to signal the boundaries of their territories; the marks may be deposited by urination, by defecation, or by rubbing parts of the bodies that bear specialised scent glands against the substrate. For example, dogs and other canids scent-mark by urination and defecation, while cats scent-mark by rubbing their faces and flanks against objects, as well as by the notoriously persistently smelly spraying of urine by tomcats. Many prosimians use territorial marking; for example, the Red-bellied Lemur creates territories for groups of two to ten individuals in the rainforests of eastern Madagascar by scent marking: the male Diademed Sifaka also scent marks defended territories in some of these same rainforests. Humans build physical impediments and visual warnings to mark defended territories. The male Western fence lizard defends a territory by posturing and combat, but less intensely after the mating season.

Invertebrates which show territoriality include some ants and bees, and the owl limpet

Spraying

Spraying (also known as territorial marking) is behaviour used by animals to identify their territory. Most commonly, this is scent marking, accomplished by depositing strong-smelling chemicals such as urine at prominent locations within the territory. Often the scent contains carrier proteins, such as the major urinary proteins, to stabilise the odors and maintain them for longer. Not only does the marking communicate to others of the same species, but it is also noted by prey species and avoided. For example felids such as leopards and jaguars mark by rubbing themselves against vegetation. Some prosimians, such as the Red-bellied Lemur, also use scent marking to establish a territory. Many ungulates, for example the Blue Wildebeest, use scent marking from two glands, the preorbital gland and a scent gland in the hoof.

Defense

Territories may be held by an individual, a mated pair, or a group. Territoriality is not a fixed property of a species: for example, robins defend territories as pairs during the breeding season and as individuals during the winter, while some nectarivores defend territories only during the mornings (when plants are richest in nectar). In species that do not form pair bonds, male and female territories are often independent, in the sense that males defend territories only against other males, and females only against other females; in this case, if the species is polygynous, one male territory will probably contain several female territories, while in some polyandrous species such as the Northern Jacana, this situation is reversed.

Quite often territories that only yield a single resource are defended. For example, European Blackbirds may defend feeding territories that are distant from their nest sites, and in some species that form leks, for example the Uganda kob (a grazing antelope), males defend the lek site (which is used only for mating).

Territoriality is only shown by a minority of species. More commonly, an individual or a group of animals will have an area that it habitually uses but does not necessarily defend; this is called its home range. The home ranges of different groups often overlap, and in the overlap areas the groups will tend to avoid each other rather than seeking to expel each other. Within the home range there may be a *core area* that no other individual group uses, but again this is as a result of avoidance rather than defense.

Behavioural ecologists have argued that food distribution determines whether a species will be territorial or not. This however, though true as far as it goes, is too narrow a point of view. As mentioned above, there are several kinds of territoriality; for example, the defence of lek areas by kob has nothing to do with food. Many other examples of territorial defence, including fish, birds or even invertebrates, are related to competition for mates or safe lairs, rather than food. Territoriality will emerge where there is a focused resource that provides enough for the individual or group, within a boundary that is small enough to be defended without the expenditure of too much effort.

Many birds, particularly seabirds, though they nest in dense communities, are nonetheless territorial in that they defend their nesting site to within the distance that they can reach while brooding. This is necessary to prevent attacks on their own chicks or nesting material from neighbours. Commonly the resulting superimposition of the short-range repulsion onto the long-range attraction characteristically leads to the well-known roughly hexagonal spacing of nests. Interestingly, one gets a similar hexagonal spacing resulting from the territorial behaviour of gardening limpets such as species of Scutellastra They vigorously defend their gardens of particular species of algae, that extend for perhaps 1–2 cm around the periphery of their shells.

Territoriality is least likely with insectivorous birds, where the food supply is plentiful but unpredictably distributed. Swifts rarely defend an area larger than the nest. Conversely, other insectivorous birds that occupy more constrained territories, such as the ground-nesting Blacksmith Lapwing may be very territorial, especially in the breeding season, where they not only threaten or attack many kinds of intruders, but have stereotyped display behaviour to deter conspecifics sharing neighbouring nesting spots.

Conversely, large solitary (or paired) carnivores, such as bears and the bigger raptors require an extensive protected area to guarantee their food supply. This territoriality will only break down when there is a glut of food, for example when Grizzly Bears are attracted to migrating salmon.

Alarm Signal

In the field of animal communication, an alarm signal is an antipredator adaptation referring to various signals emitted by social animals in response to danger. Many primates and birds have elaborate alarm calls for warning conspecifics of approaching predators. For example, the characteristic alarm call of the Blackbird is a familiar sound in many gardens. Other animals like fish and insects may use other non-auditory signals, such as chemical messages. While visual signs have been suggested as alarm signals, they are easier to pinpoint by predators and less likely to be received by conspecifics, so have tended to be treated as a signal to the predator instead.

Different calls may be used for predators on the ground or from the air. Often, the animals can tell which member of the group is making the call, so that they can disregard those of little reliability.

Evidently, alarm signals promote survival by allowing the receivers of the alarm to escape from the source of peril, but this ecological safety system may come at a cost to the signaller. When an animal calls attention to itself by signalling, it may be more likely to be eaten by a predator than if it had kept quiet. This intuition has been verified by experimental data on ground squirrel predation rates and the connection between this and the noisy chirping or whistling alarm calls. However, there is also some evidence that alarm calls can increase individual fitness as well.

Alarm calls are often high frequency sounds because these sounds are harder to localise, despite common belief.

Selective Advantage

This cost/benefit tradeoff of alarm calling behaviour has sparked many interest debates among evolutionary biologists seeking to explain the occurrence of such apparently "self-sacrificing" behaviour. The central question is this: "If the ultimate purpose of any animal behaviour is to maximise the chances that an organism's own genes are passed on, with maximum fruitfulness, to future generations, why would an individual deliberately risk destroying themselves (their entire genome) for the sake of saving others (other genomes)?".

Some scientists have used the evidence of alarm-calling behaviour to challenge the theory that "evolution works only/primarily at the level of the gene and of the gene's "interest" in passing itself along to future generations." If alarm-calling is truly an example of altruism, then our understanding of natural selection becomes more complicated than simply "survival of the fittest gene".

Other researchers, generally those who support the selfish gene theory, question the authenticity of this "altruistic" behaviour. For instance, it has been observed that Vervets sometimes emit calls in the presence of a predator, and sometimes do not. Studies show that these Vervets may call more often when they are surrounded by their own offspring and by other relatives who share many of their genes.

Other researchers have shown that some forms of alarm calling, for example, "aerial predator whistles" produced by Belding's Ground Squirrels, do not increase the chances that a caller will get eaten by a predator; the alarm call is advantageous to both caller and recipient by frightening and warding off the predator.

Another theory suggests that alarm signals function to attract further predators, which fight over the prey organism, giving it a better chance of escape. Others still suggest they are a deterrent to predators, communicating the animals alertness to the predator. One such case is the Eastern Swamphen (Porphyrio porphyrio), which gives conspicuous visual tail flicks (see also aposematism, handicap principle and stotting).

Considerable research effort continues to be directed toward the purpose and ramifications of alarm-calling behaviour, because, to the extent that this research has the ability to comment on the occurrence or non-occurrence of altruistic behaviour, we can apply these findings to our understanding of altruism in human behaviour.

Monkeys with Alarm Calls

Vervet monkeys are the typical example of both animal alarm calls and of semantic capacity in non-human animals. They have three distinct calls for leopards, snakes, and eagles, and research shows that each call elicits different responses. When vervets are on the ground they respond to the eagle alarm call by looking up and running to cover, to leopard alarm calls primarily by looking up and running into a tree, and to the snake alarm call primarily by looking down. When in trees vervets responded to the eagle alarm call by looking up and down and running out of trees, to the leopard alarm call by running higher in the tree and looking both up and down, and to the snake alarm call by looking primarily down.

Campbell's mona monkeys also generate alarm calls, but in a different way than Vervet monkeys. Instead of having discrete calls for each predator, Campbell Monkeys have two distinct types of calls which contain different calls which consist in an acoustic continuum of affixes which change meaning. It has been suggested that this is a homology to human morphology. Campbell Monkeys also demonstrated abilities similar to Vervet Monkeys' ability to distinguish likely direction of predation and appropriate responses.

That these two species use vocalisations to warn others of danger has been called by some proof of protolanguage in primates. However, there is some evidence that this behaviour does not refer to the predators themselves but to threat, distinguishing calls from words.

False Alarm Calls

Deceptive alarm calls are used by male swallows (*Hirundo rustica*). Males give these false alarm calls when females leave the nest area during the mating season, and are thus able to disrupt extra-pair copulations. As this is likely to be costly to females, it can be seen as an example of sexual conflict. Counterfeit alarm calls are also used by thrushes to avoid intraspecific competition. By sounding a bogus alarm call normally used to warn of aerial predators, they can frighten other birds away, allowing them to eat undisturbed. Vervets seem to be able to understand the referent of alarm calls instead of merely the acoustic properties, and if another species' specific alarm call (terrestrial or aerial predator, for instance) is used incorrectly with too high of a regularity, the vervet will learn to ignore the analogous vervet call as well.

Alarm Pheromones

Alarm signals need not be communicated only by auditory means. For example, many animals may use chemosensory alarm signals, communicated by chemicals known as pheromones. Minnows and catfish release alarm pheromones (Schreckstoff) when injured, which cause nearby fish to hide in dense schools near the bottom. Animals are not the only organism to communicate threats to conspecifics either; some plants are able to perform a similar trick. Lima beans release volatile chemical signals that are received by nearby plants of the same species when infested with spider mites. This 'message' allows the recipients to prepare themselves by activating defense genes, making them less vulnerable to attack, and also attracting another mite species that is a predator of spider mites (*indirect defence*). Although it is conceivable that other plants are only intercepting a message primarily functioning to attract "bodyguards", some plants spread this signal on to others themselves, suggesting an indirect benefit from increased inclusive fitness. False chemical alarm signals are also employed. The aphid *Myzus persicae* is repelled by the wild potato *Solatium berthaultii* which releases a chemical from its leaves that acts as an allomone to disrupt aphid attacks.

Meta-communication

Metacommunication is communication that indicates how verbal information should be interpreted and concerns stimuli surrounding the verbal communication that also have meaning, and which may or may not be congruent, supportive or contradictory of that verbal communication. The word was brought to prominence and perhaps coined by Gregory Bateson.

Human communication, in addition to verbal communication, involves kinesic and paralinguistic elements which can be seen as metacommunicative signals i.e. messages about messages. These indicate how the verbal communication should be understood and interpreted. Meaning does not depend only on literal verbal meaning, but is codetermined in a critical way by the intensity and inflection of the voice, facial expression, accompanying gestures, secondary signals sent to bystanders, etc. The same verbal message framed by different metacommunication can mean something entirely different, including its opposite.

In Communication: The Social Matrix of Psychiatry, Gregory Bateson defines metacommunication as "all exchanged cues and propositions about (a) codification and (b) relationship between the communicators"(Ruesch and Bateson, 1951, p. 209)

History

While metacommunication is generally viewed as originated by Gregory Bateson, Frits Staal, Emeritus Professor of Philosophy and South & Southeast Asian Studies at the University of California, Berkeley, considered that in Indian culture the concept of metalanguage originated early in the context of linguistics and speculations on language; whereas in the West, it originated late in the context of logic.

The term Kinesics, employed by Bateson, was first used (in 1952) by Ray Birdwhistell an anthropologist who wished to study how people communicate through posture, gesture, stance, and movement. Part of Birdwhistell's work involved making film of people in social situations and analysing them to show different levels of communication not clearly seen otherwise. The study was joined by several other anthropologists, including Margaret Mead and Gregory Bateson.

From 1952-1962, Bateson directed a research project on communication. This paid particular attention to logical paradoxes including Russell's paradox 1901 and to of Bertrand Russell's, Theory of Types, Russell's solution to it. Bateson and his associates here pioneered the concept of metacommunication - something that means different (often contradictory) things at different levels. Metacommunication is a characteristic feature of complex systems.

Russell's 1902 solution to his paradox comes in large part from the so-called *vicious circle principle,* that no propositional function can be defined prior to specifying the function's scope of application. In other words, before a function can be defined, one must first specify exactly those objects to which the function will apply (the function's domain). For example, before defining that the predicate "is a prime number," one first needs to define the collection of objects that might possibly satisfy the predicate, namely the set, N, of natural numbers. . It functions as a formal definition of the function of metacommunication in communication.

Applications

Ivan Pavlov: In Ivan Pavlov's experiment in which dogs where trained to salivate upon hearing a bell ring. This was accomplished by ringing a bell just prior to feeding the dogs. After repeating this procedure for some time it was found that the dogs would salivate after hearing the bell — without the need for food being presented. They had learned that the ringing of the bell communicated "food is on the way."

Something that is not often discussed in context with this experiment is the fact that the dogs would not salivate unless they were wearing a special harness. When exposed to the bell ringing without wearing the harness, the dogs did not salivate. The dogs only salivated upon hearing the bell *while wearing the harness*. The bell ringing was direct communication of information. The context of the communication (the metacommunication) also conveyed information.

Mating rituals: signals made by members of one sex to attract or maintain the attention of potential mate, or to cement a pair bond. These frequently involve the display of body parts, body postures (gazelles assume characteristic poses as a signal to initiate mating), or the emission of scents or calls, that are unique to the species, thus allowing the individuals to avoid mating with members of another species which would be infertile. Animals that form lasting pair bonds often have symmetrical displays that they make to each other: famous examples are the mutual presentation of reeds by Great Crested Grebes, studied by Julian Huxley, the *triumph displays* shown by many species of geese and penguins on their nest sites and the spectacular courtship displays by birds of paradise and manakins.

Food-related signals: many animals make "food calls" that attract a mate, or offspring, or members of a social group generally to a food source. When parents are feeding offspring, the offspring often have begging responses (particularly when there are many offspring in a clutch or litter - this is well known in altricial songbirds, for example). Perhaps the most elaborate food-related signal is the dance language of honeybees studied by Karl von Frisch. Young ravens signal to older, more experienced ravens when they come across new or untested food.

Interpretation of Animal Communication

It is important to note that whilst many gestures and actions have common, stereotypical meanings, researchers regularly seem to find that animal communication is often more complex and subtle than previously believed, and that the same gesture may have multiple distinct meanings depending on context and other behaviours. So generalisations such as "X means Y" are *often*, but not *always* accurate. For example, even a simple domestic dog's tail wag may be used in subtly different ways to convey many meanings including:

- Excitement
- Anticipation
- Playfulness
- Contentment/enjoyment
- Relaxation or anxiety
- Questioning another animal or a human as to intentions
- Tentative role assessment on meeting another animal
- Reassurance ("I'm hoping to be friendly, are you?")
- Brief acknowledgement ("I hear you", or "I'm aware and responsive if you want my attention")
- Statement of interest ("I want that (food/toy/activity), if you're willing")
- Uncertainty/apprehension
- Submissive placation (if worried by a more dominant animal).

Combined with other body language, in a specific context, many gestures such as yawns, direction of vision, and so on all convey meaning. Thus statements that a particular action "means" something should always be interpreted to mean "often means" something. As with human beings, who may smile or hug or stand a particular way for multiple reasons, many animals reuse gestures too.

Intraspecies vs. Interspecies Communication

The sender and receiver of a communication may be of the same species or of different species. The majority of animal communication is intraspecific (between two or more individuals of the same species). However, there are some important instances of interspecific communication. Also, the possibility of interspecific communication, and the form it takes, is an important test of some theoretical models of animal communication.

Intraspecies Communication

The majority of animal communication occurs within a single species, and this is the context in which it has been most intensively studied.

Most of the forms and functions of communication described above are relevant to intra-species communication.

Interspecies Communication

Many examples of communication take place between members of different species. Animals communicate to other animals with various signs: visual, sound, echolocation, body language, and smell.

Prey to Predator

If a prey animal moves or makes a noise in such a way that a predator can detect and capture it, that fits the definition of "communication" given above. This type of communication is known as interceptive eavesdropping, where a predator intercepts the message being conveyed to conspecifics.

There are however some actions of prey species that are clearly communications to actual or potential predators. A good example is warning colouration: species such as wasps that are capable of harming potential predators are often brightly coloured, and this modifies the behaviour of the predator, who either instinctively or as the result of experience will avoid attacking such an animal. Some forms of mimicry fall in the same category: for example hoverflies are coloured in the same way as wasps, and although they are unable to sting, the strong avoidance of wasps by predators gives the hoverfly some protection.

There are also behavioural changes that act in a similar way to warning colouration. For example, canines such as wolves and coyotes may adopt an aggressive posture, such as growling with their teeth bared, to indicate they will fight if necessary, and rattlesnakes use their well-known rattle to warn potential predators of their poisonous bite. Sometimes, a behavioural change and warning colouration will be combined, as in certain species of amphibians which have a brightly coloured belly, but on which the rest of their body is coloured to blend in with their surroundings. When confronted with a potential threat, they show their belly, indicating that they are poisonous in some way.

Another example of prey to predator communication, is referred to as a pursuit-deterrent signal. Pursuit-deterrent signals occur when prey indicates to a predator that pursuit would be unproûtable because the signaler is prepared to escape.

Pursuit-deterrent signals provide a beneût to both the signaler and receiver; they prevent the sender from wasting time and energy ûeeing, and they prevent the receiver from investing in a costly pursuit that is unlikely to result in capture. Such signals can advertise prey's ability to escape, and reûect phenotypic condition (quality advertisement), or can advertise that the prey has detected the predator (perception advertisement).

Pursuit-deterrent signals have been reported for a wide variety of taxa, including ûsh (Godin and Davis 1995), lizards (Cooper et al. 2004), ungulates (Caro 1995), rabbits (Holley 1993), primates (Zuberbuhler et al. 1997), rodents (Shelley and Blumstein 2005, Clark 2005), and birds (Alvarez 1993, Murphy 2006, 2007). The most familiar example of quality advertisement pursuit-deterrent signal is *stotting*, a pronounced combination of running while simultaneously hopping shown by some antelopes such as Thomson's gazelle in the presence of a predator. At least 11 hypothesis for stotting have been proposed. A leading theory today is that it alerts predators that the element of surprise has been lost. Predators like cheetahs rely on surprise attacks, proven by the fact that chases are rarely successful when they stot. Predators know not to waste energy on a chase that will likely be unsuccessful (optimal foraging behaviour).

Predator to Prey

Some predators communicate to prey in ways that change their behaviour and make them easier to catch, in effect deceiving them. A well-known example is the angler fish, which has a fleshy growth protruding from its forehead and dangling in front of its jaws; smaller fish try to take the lure, and in so doing are perfectly placed for the angler fish to eat them.

Human/Animal Communication

Various ways in which humans interpret the behaviour of domestic animals, or give commands to them, fit the definition of interspecies communication. Depending on the context, they might be considered to be predator to prey communication, or to reflect forms of commensalism.

The recent experiments on animal language are perhaps the most sophisticated attempt yet to establish human/animal communication, though their relation to natural animal communication is uncertain. Lacking in the study of human-animal communication is a focus on expressive communication from animal to human specifically. Other than a few natural expressions animals (especially dogs) use to communicate to humans, scientists in general do not pursue expanding the expressive/productive communication of domesticated animals.

Horses are taught to not communicate (for safety). Dogs and horses are generally not encouraged to communicate expressively, but are encouraged to develop receptive language (understanding). One scientist, Sean Senechal has pursued (since the late 1990's) developing, studying, and using the learned visible, expressive language in dogs and horses. By teaching these animals a gestural (human made) ASL-like language animals have been found to learn and use the new signs on their own to get what they need. Senechal's book Dogs Can Sign, Too documents this process.

Other Aspects of Animal Communication

Evolution of Communication

The importance of communication is clear from the fact that animals have evolved elaborate body parts to facilitate it. They include some of the most striking structures in the animal kingdom, such as the peacock's tail. Birdsong appears to have brain structures entirely devoted to its production. But even the red spot on a herring gull's bill, and the modest but characteristic bowing behaviour that displays it, require evolutionary explanation.

There are two aspects to the required explanation:

- identifying a route by which an animal that lacked the relevant feature or behaviour could acquire it;
- identifying the selective pressure that makes it adaptive for animals to develop structures that facilitate communication, emit communications, and respond to them.

Significant contributions to the first of these problems were made by Konrad Lorenz and other early ethologists. By comparing related species within groups, they showed that movements and body parts that in the primitive forms had no communicative function could be "captured" in a context where communication would be functional for one or both partners, and could evolve into a more elaborate, specialised form. For example, Desmond Morris showed in a study of grass finches that a beak-wiping response occurred in a range of species, serving a preening function, but that in some species this had been elaborated into a courtship signal.

The second problem has been more controversial. The early ethologists assumed that communication occurred for the good of the species as a whole, but this would require a process of group selection which is believed to be mathematically impossible in the evolution of sexually reproducing animals. Altruism towards an unrelated group is not widely accepted in the scientific community, but rather can be seen as a sort of reciprocal altruism, expecting the same behaviour from others, a benefit of living in a group. Sociobiologists argued that behaviours that benefited a whole group of animals might emerge as a result of selection pressures acting solely on the individual. A gene-centered view of evolution proposes that behaviours that enabled a gene to become wider established within a population would become positively selected for, even if their effect on individuals or the species as a whole was detrimental.

In the case of communication, an important discussion by John Krebs and Richard Dawkins established hypotheses for the evolution of such apparently altruistic or mutualistic communications as alarm calls and courtship signals to emerge under individual selection. This led to the realisation that communication might not always be "honest" (indeed, there are some obvious examples where it is not, as in mimicry). The possibility of evolutionarily stable dishonest communication has been the subject of much controversy, with Amotz Zahavi in particular arguing that it cannot exist in the long term. Sociobiologists have also been concerned with the evolution of apparently excessive signalling structures such as the peacock's tail; it is widely thought that these can only emerge as a result of sexual selection, which can create a positive feedback process that leads to the rapid exaggeration of a characteristic that confers an advantage in a competitive mate-selection situation.

One theory to explain the evolution of traits like a peacock's tail is 'runaway selection'. This requires two traits-a trait that exists, like the bright tail, and a prexisting bias in the female to select for that trait. Females prefer the more elaborate tails, and thus those males are able to mate successfully. Exploiting the psychology of the female, a positive feedback loop is enacted and the tail becomes bigger and brighter. Eventually, the evolution will level off because the survival costs to the male do not allow for the trait to be elaborated any further. Two theories exist to explain runaway selection. The first is the good genes hypothesis. This theory states that an elaborate display is an honest signal of fitness and truly is a better mate. The second is the handicap hypothesis. This explains that the peacock's tail is a handicap, requiring energy to keep and makes it more visible to predators. Regardless, the individual is able to survive, even though its genes are not as good per se.

Cognitive Aspects

Ethologists and sociobiologists have characteristically analysed animal communication in terms of more or less automatic responses to stimuli, without raising the question of whether the animals concerned understand the meaning of the signals they emit and receive. That is a key question in animal cognition. There are some signalling systems that seem to demand a more advanced understanding. A much discussed example is the use of alarm calls by vervet monkeys. Robert Seyfarth and Dorothy Cheney showed that these animals emit different alarm calls in the presence of different predators (leopards, eagles, and snakes), and the monkeys that hear the calls respond appropriately - but that this ability develops over time, and also takes into account the experience of the individual emitting the call. Metacommunication, discussed above, also seems to require a more sophisticated cognitive process.

A recently published paper demonstrated that bottlenose dolphins can recognise identity information from whistles even when otherwise stripped of the characteristics of the whistle; making dolphins the only animals other than humans that have been shown to transmit identity information independent of the caller's voice or location. The paper concludes that:

> *The fact that signature whistle shape carries identity information independent from voice features presents the possibility to use these whistles as referential signals, either addressing individuals or referring to them, similar to the use of names in humans. Given the cognitive abilities of bottlenose dolphins, their vocal learning and copying skills, and their fission–fusion social structure, this possibility is an intriguing one that demands further investigation. —V. M. Janik,* et al.

Animal Communication and Human Behaviour

Another controversial issue is the extent to which humans have behaviours that resemble animal communication, or whether all such communication has disappeared as a result of our linguistic capacity. Some of our bodily features - eyebrows, beards and moustaches, deep adult male voices, perhaps female breasts - strongly resemble adaptations to producing signals. Ethologists such as Irenäus Eibl-Eibesfeldt have argued that facial gestures such as smiling, grimacing, and the *eyebrow flash* on greeting are universal human communicative signals that can be related to corresponding signals in other primates. Given the recency with which spoken language has emerged, it is very likely that human body language does include some more or less involuntary responses that have a similar origin to the communication we see in other animals.

Humans also often seek to mimic animals' communicative signals in order to interact with the animals. For example, cats have a mild affiliative response involving closing their eyes; humans often close their eyes towards a pet cat to establish a tolerant relationship. Stroking, petting and rubbing pet animals are all actions that probably work through their natural patterns of interspecific communication.

Dogs have shown an ability to understand communication from a species other than their own. They were able to use human communicative gestures such as pointing and looking to find hidden food and toys.

A new approach in the 21 century of studying animal communication uses applied behavioural analysis (ABA), specifically Functional Communication Training (FCT). This FCT previously has been used in schools and clinics with humans with special needs, such as children with autism, to help them develop language. Sean Senechal, at the AnimalSign Center has been using an approach similar to this FCT with domesticated animals, such as dogs (since 2004) and horses (since 2000) with encouraging results and benefits to the animals and people. Functional communication training for animals, Senechal calls AnimalSign Language. This includes teaching communication through gestures (like simplified ASL), pictures (PECS), tapping, and vocalisation. The process for animals includes simplified and modified techniques.

Human–animal Communication

Human–animal communication is the communication observed between humans and other animals, from non-verbal cues and vocalisations through to, potentially, the use of a sophisticated language.

Human–animal communication is easily observed in everyday life. The interactions between pets and their owners, for example, reflect a form of spoken, while not necessarily verbal dialogue. A dog being scolded does not need to understand every word of its admonishment, but is able to grasp the message by interpreting cues such as the owner's stance, tone of voice, and body language. This communication is two-way, as owners can learn to discern the subtle differences between barks and meows ... one hardly has to be a professional animal trainer to tell the difference between the bark of an angry dog defending its home and the happy bark of the same animal while playing. Communication (often nonverbal) is also significant in equestrian activities such as dressage.

Word Repetition in Birds

Although the word repetition skills observed in some birds (most famously parrots) should not be mistaken for lingual communication, this tendency has nonetheless influenced fictional portrayals of animal communication, as sentient talking parrots and similar birds are common in children's fiction, such as the talking, loud-mouth parrot Iago of Disney's *Aladdin*. Bruce Thomas Boehner's book *Parrot Culture: Our 2,500-Year-Long Fascination with the World's Most Talkative Bird* explores this issue thoroughly.

The Next Level: Language

Achieving a deeper level of communication between animals and humans has long been a goal of science. Perhaps the most famous example of recent decades has been Koko, a gorilla who is supposedly able to communicate with humans using a system based on American Sign Language with a "vocabulary" of over 1000 words.

John Lilly and Cetacean Communication

In the 1960s, John Lilly, M.D., prolific writer and explorer of consciousness via the isolation tank (his invention), and contemporary and associate of Timothy Leary, began experiments in the Virgin Islands aiming to establish meaningful communication between humans and the bottlenose dolphin (*Tursiops truncatus*). Lilly financed, mostly personally, a human-dolphin cohabitat, a house on the ocean's shore that contained an area that was partially flooded and allowed a human and dolphin to live together in the same space, sharing meals, play, language lessons, and even sleep.

Two experiments of this sort are explained in detail in Lilly's popular books (see John Lilly for bibliography). The first experiment was more of a test run to check psychological and other strains on the human and cetacean participants, determining the extent of the need for other human contact, dry clothing, time alone, and so on. Despite tensions after several weeks, the experimenter, Margaret C. Howe, agreed to a two-and-a-half month experiment, living isolated with 'Peter' dolphin.

A basic outline of Peter dolphin's linguistic progress is as follows: early lessons involved mostly noise and interruptions from Peter during English lessons, and a food reward of fish was necessary for him to 'attend class.' After several weeks, a concerted effort by Peter to imitate the instructor's speech was evident, and human-like sounds were apparent, and recorded. More interesting was the dolphin's immediate grasp of basic semantics, such as the different aural indicators for 'ball' and 'doll' and other toys present in the aquarium. Peter was able to perform tasks such as retrieval on the (aurally) indicated object without fail. Later in the project the dolphin's ability to process linguistic syntax was made apparent, in that Peter could distinguish between the commands (e.g., only) "Bring the ball to the doll," and "Bring the doll to the ball." This ability not only demonstrates the bottlenose dolphin's grasp of basic grammar, but also implies the dolphins' own language must include some such syntactical rules. The correlation between length and 'syllables' (bursts of the dolphin's sound) with the instructor's speech also went from essentially zero at the beginning of the session to almost a perfect correlation by its completion. I.e., a sentence spoken by the instructor involving 35 syllables and lasting 8 seconds would be met with an 8-second burst of sound from Peter dolphin involving 35 easily-discernible 'syllables' or bursts of sound.

Much later, experiments by Louis Herman, a former collaborator and student of Lilly's, demonstrated the crossmodal perceptual ability of dolphins. Dolphins typically perceive their environment through sound waves generated in the melon of their skulls, through a process known as echolocation (similar to that seen in bats, though the mechanism of production is different). The dolphin's eyesight however is also fairly good, even by human standards, and Herman's research found that any object, even of complex and arbitrary shape, identified either by sight or sound by the dolphin, could later be correctly identified by the dolphin with the alternate sense modality with almost 100 per cent accuracy, in what is classically known in psychology and behaviourism as a match-to-sample test. The only errors noted were presumed to have been a misunderstanding of the task during the first few trials, and not an inability of the dolphin's perceptual apparatus. This capacity is strong evidence for abstract and conceptual thought in the dolphin's brain, wherein an idea of the object is stored and understood not merely by its sensory properties; such abstraction may be argued to be of the same kind as complex language, mathematics, and art, and implies a potentially very great intelligence and conceptual understanding within the brains of tursiops and possibly many other cetaceans. Accordingly, Lilly's interest later shifted to whale song and the possibility of high intelligence in the brains of large whales, and Louis Herman's research at the now misnomered Dolphin Institute in Honolulu, Hawaii, focuses exclusively on the Humpback whale.

Animal Communication as Entertainment

Though animal communication has always been a topic of public comment and attention, for a period in history it surpassed this and became sensational popular entertainment. From the late 18th century through the mid 19th century, a succession of "learned pigs" and various other animals were displayed to the public in for-profit performances, boasting the ability to communicate with their owners (often in more than one language), write, solve math problems, and the like. One poster dated 1817 shows a group of "Java sparrows" who are advertised as knowing *seven languages*, including Chinese and Russian. One pig of the era was so famous that it performed for royalty, and an obituary upon its death claimed that it made more money than any actor or actress of the same time; a fact that, whether strictly true, was at least believable to contemporary readers. By the late 1840s the fad had died down considerably. While the occasional appearance by a "learned" animal continued into the radio and television eras, it was by then generally understood that feats such as using cards to spell words, barking or tapping a hoof to solve equations, and the like were the products of training rather than actual *communication*. Though the tradition continues to this day on the "Stupid Pet Tricks" segment of Late Night with David Letterman, it seems likely that the era of trained pigs entertaining the crowned heads of Europe are over.

BowLingual

One real-world example of a technological means of one-way human–animal communication is BowLingual, a Japanese device which claims to translate barks from dozens of different breeds of dogs, including mixed-breeds. Based largely on Dr. Matsumi Suzuki's Animal Emotion Analysis System developed at Japan Acoustic Laboratory, the device outputs one of 200 phrases (grouped into six different moods), supposedly reflecting "meaning" of the dog's bark. The device was apparently successful enough in Japan to be brought to the American market, and was even named one of 2002's best inventions by Time Magazine. However, reports of the BowLingual's accuracy have been mixed at best, with popular product-review website Epinions giving it a low 1.5 stars average.

Human-animal Communication in Culture

The concept of human–animal communication has existed in culture for longer than recorded history, being an element of many myths and folk tales of numerous cultures, and continues in modern popular entertainment. This section lists some examples of this, divided by the method of communication (magical/supernatural, innate natural ability, technological, and unspecified/misc). Topics that are beyond the scope of this article and will not be listed here include intraspecies communication (e.g. *Watership Down*) and interspecies communication not involving humans (e.g. *Redwall*) or sentient animals who can think but not communicate with humans (e.g. *Garfield*; Buck of Married... with Children; Blondi of Kingdom Hospital).

Magic and Supernatural

- In many fantasy role playing games, Druid characters are able to speak with animals through the use of a spell. The Dungeons & Dragons version of this spell is called "Speak with Animals".
- Eliza Thornberry of Nickelodeon's animated TV series The Wild Thornberrys can speak with animals after a spell is placed on her by an African tribal shaman.
- In the Harry Potter series of books, Harry is a parselmouth (able to speak with snakes in their own language, parseltongue, which sounds like hissing and spitting to the ears of those without this (apparently hereditary) skill).
- In The Immortals series of books, Veralidaine Sarrasri (Daine) has "wild magic" which enables her to communicate with animals. After being trained in the use of her wild magic, she learns to enter the minds of animals and shapeshift.
- In the Obernewtyn Chronicles, some of the characters have the ability to mentally communicate with animals, an ability known as 'beastspeaking'. All animals have the same 'language', which comprises of mental pictures.

Innate Ability

- Cypher, real name Douglas Ramsey, of Marvel Comics' The New Mutants, has a "mutant" ability to instantly translate any language he hears or sees, including animal languages.
- Doctor Dolittle, subject of a series of children's books by Hugh Lofting as well as various film and stage adaptions, is a doctor whose ability to speak with animals makes him successful in dealing with animals but closes him off from most humans.

Technological

- In Star Trek, the communicator badges worn by Starfleet crew members allow direct translation between humans and various sentient and semi-sentient aliens and creatures, though it is undetermined whether they work with "dumb" animals. It is presumed that this is not possible, as the relatively few domesticated animals seen on the show do not appear to converse with their owners (Data's cat, and Archer's beagle, for example). The council-chamber of the Xindi accommodates her Aquatic members in a water-tank.

- In issue 285 of DC's Detective Comics, Martian Manhunter fights a brigade of "Martian mandrills" which were being manipulated by villains through the use of a communicator device stolen from the mandrills' rocket.
- On the television cartoon Krypto the Superdog, Krypto has an "intergalactic communicator" device in his dog tag, which allows Kevin Whitney to communicate with him. The device is technology from Krypton, Superman's homeworld.
- In the anime, Immortal Grand Prix (IGPX), Team Satomi's Midfielder, Amy, uses a machine to communicate with her cat, Luca, while racing with him in her mech. The same can be said for Team Edgeraid's Forward, Bjorn, and his dog, Sola.

Unspecified and Miscellaneous

- In Hergé's *Tintin* comics, Tintin's dog Snowy is sentient and able to "think". Although his thoughts are written in word bubbles rather than thought bubbles, it is generally assumed that the human characters cannot understand him. Once, in the early volume *Tintin in America*, Tintin was able to directly understand Snowy. Hergé did not elaborate on why this was so; presumably it was used simply to advance the plot and not to bog down the story with a "talking dog" element.
- The American television show Mr. Ed centers around a horse's ability to communicate with his owner, Wilbur. The plot of this television series was inspired by the movie character Francis the Talking Mule.

Animal Language

Animal language is the modeling of human language in non human animal systems. While the term is widely used, researchers agree that animal languages are not as complex or expressive as human language.

Some researchers including the linguist Charles Hockett, who proposed a list of design features of Human Language, argue that there are significant differences separating human language from animal communication even at its most complex, and that the underlying principles are not related. Accordingly, Thomas A. Sebeok has proposed not to use the term 'language' in case of animal sign systems.

Others argue that an evolutionary continuum exists between the communication methods these animals use and human language. Examining this continuum could help explain how humanity evolved its incredibly sophisticated proficiency for language.

Aspects of Human Language

The following properties of human language have been argued to separate it from animal communication:

- *Arbitrariness:* There is not necessarily a rational relationship between a sound or sign and its meaning. (There is nothing intrinsically "housy" about the word "house". i.e. *symbolism*)
- *Cultural transmission:* Language is passed from one language user to the next, consciously or unconsciously.
- *Discreteness:* Language is composed of discrete units that are used in combination to create meaning.
- *Displacement:* Languages can be used to communicate ideas about things that are not in the immediate vicinity either spatially or temporally, or both.
- *Duality:* Language works on two levels at once, a surface level and a semantic (meaningful) level.
- *Metalinguistics:* Ability to discuss language itself.
- *Productivity:* A finite number of units can be used to create an indefinitely large number of utterances.

Research with apes, like that of Francine Patterson with Koko or Herbert Terrace with Nim Chimpsky, suggested that apes are capable of using language that meets some of these requirements such as arbitrariness, cultural transmission, discreteness and productivity. However, no experiment has yet shown a non-human being to be proficient in all of these areas.

In the wild chimpanzees have been seen "talking" to each other, when warning about approaching danger. For example, if one chimpanzee sees a snake, he makes a low, rumbling noise, signalling for all the other chimps to climb into nearby trees. In this case, the chimpanzees' communication is entirely contained to an observable event, demonstrating a lack of displacement.

Arbitrariness has been noted in meerkat calls; bee dances show elements of spatial displacement; and cultural transmission has possibly occurred between the celebrated bonobos Kanzi and Panbanisha.

Human language may not be completely "arbitrary". Some research has shown that almost all humans naturally demonstrate limited crossmodal perception (e.g. synesthesia), as illustrated by the Kiki and Booba study.

Claims that animals have language skills akin to humans however, are extremely controversial. As Pinker illustrates in his book the "The

Language Instinct", claims that chimpanzees can acquire language are exaggerated and rest on very limited or specious data.

Bee Learning and Communication

Honey bees learn and communicate in order to find food sources and for other means.

Figure: *Swarming bees require good communication in order to congregate all in the same spot*

Learning

Learning is essential for efficient foraging. Honey bees are unlikely to make many repeat visits if a plant provides little in the way of reward. A single forager will visit different flowers in the morning and, if there is sufficient attraction and reward in a particular kind of flower, she will make visits to that type of flower for most of the day, unless the plants stop producing reward or weather conditions change. Honey bees are quite adept at associative learning, and many of the standard phenomena of classical conditioning take the same form in honey bees as they do in the vertebrates that are the more usual subjects of such experiments.

Foragers were trained to enter a simple Y-shaped maze that had been marked at the entrance with a particular colour. Inside the maze was a branching point where the bee was required to choose between two paths. One path, which led to the food reward, was marked with the same colour that had been used at the entrance to the maze, while the other was marked with a different colour. Foragers learned to choose the correct path, and continued to do so when a different kind of marker (black and white stripes oriented in various directions) was substituted for the coloured markers. When the experimental conditions were reversed, rewarding bees for choosing the inner passage marked with a symbol that was different from the entrance symbol, the bees again learned to choose the correct path. Extending the length of the tunnel to increase the time between seeing the one marker indicating the correct path and a second marker identifying the correct path show that the bees can retain the information in their visual working memory for about 5 seconds, equivalent to the short-term memory of birds.

Colour Learning in Honeybees

One of the most common ways that honey bees, *Apis mellifera*, demonstrate associative learning is in the context of colour recognition and discrimination tasks. Just as vertebrate species such as mice or pigeons that can be trained to perform associative learning tasks, honey bees make excellent subjects for tasks involving discrimination and colour memory. Beginning in the early 1900s, scientists Karl von Frisch and later Randolf Menzel began asking questions about the existence, learning rates, memory, and timing of colour vision in bees.

Colour Discrimination

The Austrian zoologist Karl von Frisch began the exploration of colour vision in honey bees when he asked the first question in 1919: does colour vision in bees exist? By making use of bees associative learning abilities he performed an elegant experiment to show that honey bees were in fact capable of colour discrimination.

To test colour vision, von Frisch first trained his honey bees to feed from a small dish filled with a nectar-like sugar water. This dish was placed over a small piece of blue coloured cardboard so that the colour was visible to the bees as they fed. Once the bees had become accustomed to the blue cardboard, von Frisch surrounded the blue piece of cardboard with other identically sized pieces in varying shades of gray and placed small dishes over each piece. If bees could not discriminate between colours, they would be unable to distinguish the blue piece from the many gray toned pieces. In the case that bees did not have colour vision then, von Frisch predicted that the bees would visit the gray and blue pieces with equal frequency, as they would not be able to tell a difference between them.

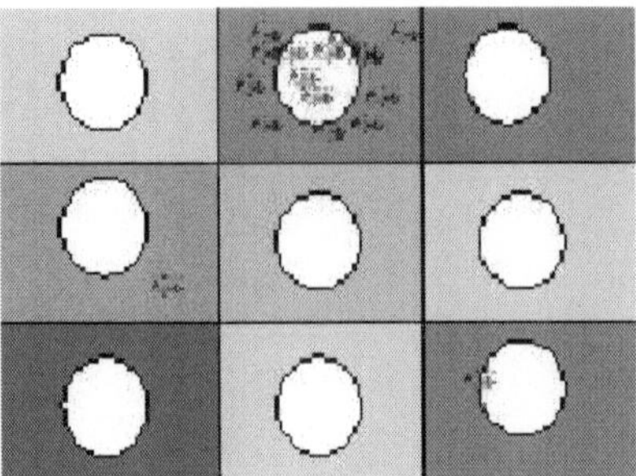

Figure: *Testing for colour vision in honey bees. The majority of bees flew directly to the dish with the blue background as they had been trained to do. Thus, they were able to discriminate between gray and blue backgrounds, showing their capability for colour vision.*

When he allowed bees access to the dishes, however, he found that the vast majority of the bees flew directly to the blue piece of cardboard on which they had previously obtained their sugar-water reward. The bees largely ignored the gray pieces which had not been rewarded. This directed exploration and targeting of the blue cardboard showed the bees could indeed discriminate between the gray and blue shades, showing that bees do possess colour vision. Von Frisch repeated this same basic experiment to show that bees produced the same results with other colours like red and yellow. Later other researchers were able to apply this excellent experimental design to other vertebrates as well, making it an invaluable insight into testing colour vision in many organisms.

Colour Learning Rates and Preferences

After von Frisch's initial studies, the German scientist Randolf Menzel continued the study of colour vision in honey bees and performed more detailed tests. He was curious about which colours honey bees would be able to learn fastest and whether or not bees had a greater aptitude for learning certain colours.

He used lights of varying colour and intensity to illuminate circles of light on a solid surface. This set up was similar to the pieces of coloured cardboard employed by von Frisch, but by using light instead of cardboard, Menzel was able to change the intensity and colour of light easily. He could simply adjust the projection of the light to create a wide variety of different experimental set-ups.

To test the intricacies of the bee colour vision von Frisch had established, Menzel performed an experiment that aimed to test bees ability to distinguish between two different colours. To do this, Menzel used a projected circle of coloured light surrounding a small dish that could hold a sugar-water reward. Menzel then projected a second circle of differently coloured light surrounding a second dish some distance away from the first.

Figure: *Honey bee collecting pollen*

Next, a single bee was placed equidistant between these two different lights and allowed to choose which dish to search for a sugar-water treat. Only one of the coloured light circles surrounded a dish that contained sugar-water; the other was empty. Menzel was then able to measure how quickly the bees learned to preferentially search only the rewarded light and to ignore the dish surrounded by unrewarded light.

The results of the experiment showed that bees did not learn to discriminate between all colour pairs equally well. The fastest rate of learning was when violet light was rewarded. The colour that the bees had the most difficulty learning was green, and all other colours fell somewhere in between. This evidence of inherent bias is evolutionarily reasonable given that colour vision in bees allows them to distinguish between different nectar-bearing flowers, much like the rewarded dishes. As more flowers are purple than green it makes sense that bees would be more sensitive to colours likely to result in nourishment.

Colour Memory

After this work on colour preferences, Menzel extended his experiments to study memory in honey bee colour vision. He wanted to know how many trials were necessary for honey bees to reliably choose a previously rewarded colour when presented with several choices for potential rewards and how long honey bees could retain information about which colour would be rewarded.

To test these questions, Menzel performed a variety of experiments. First, he presented individual bees with a sugar reward on a coloured background for just a single trial. He then kept these bees in small cages for several days without any further trials. After a few days, he presented each bee with several dishes each on a different coloured background at once. One of the colours was the same as that used during the initial trial. The others were novel, unrewarded colours. Amazingly, after just one trial and several days without any exposure to the rewarded colour, bees correctly chose to explore the colour used in the first trial more than fifty-percent of the time.

Menzel then repeated this experiment with another group of bees, keeping all factors the same except that in the second round of testing he gave the bees three initial trials with the rewarded colour instead of just one. After several days in confinement when the bees were presented with a choice of colours just as in the first experiment, they virtually always chose the colour that had been used during the first three trials.

This ability to retain information about colour-linked rewards over a period of several days and after only very minimal exposure to the coloured background indicates the great strength of honeybees memory with respect to colour vision.

Timing in Colour Learning

One of von Frisch's students, Elizabeth Opfinger, observed that bees would learn colour when approaching a feeder. Menzel took this question further: when do bees register and learn colour? He wanted to know if bees registered colour before, during, or after receiving their sugar-water reward. In order to explore this intriguing question, Menzel displayed the colour beneath a rewarded dish at different stages of the honey bee feeding process: during approach, feeding and departure.

The outcome of this experiment revealed that bees register colour during both the approach and feeding stages of the exposure process. In order for a bee to accurately remember a given colour, it must be present for approximately five seconds in total. Although it varies slightly, Menzel and his colleagues found that bees usually remember best when the stimulus is present for about three seconds during the approach and two seconds after landing and beginning to feed.

The Neurobiology of Colour Vision

Colour vision in honey bees can also be examined from a neurobiological perspective in terms of the structure and organisation of their compound eyes.

In 1975 Menzel published a seminal paper describing the morphology and spectral sensitivity of the honey bee eye. He examined colour coding the honey bee retina by using a technique to mark individual cells with a fluorescent dye and record from these cells as single units. Such fine structure analysis allowed him to determine that there are three types of receptors in the honey bee eye: 1) UV receptors, 2) blue receptors, and 3) green receptors. The three receptors are dominated by three rhodopsin-like pigments. These pigments have maximal absorbance at wavelengths corresponding to 350 nm, 440 nm, and 540 nm.

As the cells were examined in detail, certain features were distinguishable for each type of receptor cell. UV cells were found to form the longest visual fibres. These long visual fibers penetrated the lamina with arborisations, a tree-like branching of the fibers and spines. Blue and green receptor cells have more shallow fibers. Interestingly, Menzel found that most of the cells he studied had secondary sensitivities that corresponded to wavelength regions at which the other two receptor types were maximally active. He used spectral efficiency experiments to show that such corresponding wavelength receptivity is the result of electric coupling.

Communication

Foragers communicate their floral findings in order to recruit other worker bees of the hive to forage in the same area. The factors that determine recruiting success are not completely known but probably include evaluations of the quality of nectar and/or pollen brought in.

There are two main hypotheses to explain how foragers recruit other workers — the "waggle dance" or "dance language" theory and the "odor plume" theory. The dance language theory is far more widely accepted, and has far more empirical support. The theories also differ in that the former allows for an important role of odor in recruitment (i.e., effective recruitment relies on dance *plus* odor), while the latter claims that the dance is essentially irrelevant (recruitment relies on odor alone).

It has long been known that successfully foraging Western honey bees perform a dance on their return to the hive, known as *waggle dance*, indicating that food is farther away, while the *round dance* is a short version of the *waggle dance*, indicating that food is nearby. The laden forager dances on the comb in a circular pattern, occasionally crossing the circle in a zig-zag or waggle pattern. Aristotle described this behaviour in his Historia Animalium. It was thought to attract the attention of other bees.

In 1947, Karl von Frisch correlated the runs and turns of the dance to the distance and direction of the food source from the hive. The orientation of the dance correlates to the relative position of the sun to the food source, and the length of the waggle portion of the run is correlated to the distance from the hive. Also, the more vigorous the display is, the better the food. There is no evidence that this form of communication depends on individual learning.

Von Frisch performed a series of experiments to validate his theory. He was awarded the Nobel Prize in Physiology or Medicine in 1973 for his discoveries.

One of the most important lines of evidence on the origin and utility of the dance is that all of the known species and races of honey bees exhibit the behaviour, but details of its execution vary among the different species. For example, in *Apis florea* and *Apis andreniformis* (the "dwarf honeybees") the dance is performed on the dorsal, horizontal portion of the nest, which is exposed. The runs and dances point directly toward the resource in these species. Each honey bee species has a characteristically different correlation of "waggling" to distance, as well. Such species-specific behaviour suggests that this form of communication does not depend on learning but is rather determined genetically. It also suggests how the dance may have evolved.

Various experiments document that changes in the conditions under which the dance is performed lead to characteristic changes in recruitment to external resources, in a manner consistent with von Frisch's original conclusions. Researchers have also discovered other forms of honeybee dance communication, such as the tremble dance.

Odor Plume

While the majority of researchers believe that bee dances give enough information to locate resources, proponents of the odor plume theory argue that the dance gives no actual guidance to a nectar source. They argue that bees instead are primarily recruited by odor. The purpose of the dance is simply to gain attention to the returning worker bee so she can share the odor of the nectar with other workers who will then follow the odor trail to the source.

The primary lines of evidence used by the odor plume advocates are:

1. clinical experiments with odorless sugar sources which show that worker bees are unable to recruit to those sources and
2. logical difficulties of a small-scale dance (a few centimeters across) giving directions precise enough to hold the other bees

on course during a flight that could be several kilometers long. Misreading by even a few degrees would lead the bee off course by hundreds of meters at the far end.

Neither of these points invalidate the dance theory, but simply suggest out that odor might be involved, which is indeed conceded by all proponents of dance theory. Critics of the odor plume theory counter that most natural nectar sources are relatively large - orchards or entire fields. Precision may not be necessary or even desirable. They have also challenged the reproducibility of the odorless source experiment.

Significant to the argument are the elegant experiments of William F. Towne, of the Kutztown University in Pennsylvania, such as this pdf file, in which hives are moved to "mirror image" terrain settings, and thus fooled into both dancing about the wrong location for a nectar source, and successfully recruiting foragers to that wrong location, but only when the sun is obscured by clouds, forcing them to rely on terrain-based navigation rather than "solar ephemeris" based navigation. As the cloud cover breaks up, more and more bees correct their dances to indicate the actual location of nectar, and forager visits shift to the correct location.

The academic debate between these two theories is extremely polarised and often hostile. See here for an overview. Adrian Wenner, a modern bee researcher, is the chief proponent of the odor plume theory (anti-dance). One supporter of Wenner's theories, Julian O'Dea, has proposed an evolutionary explanation for the "waggle dance" that does not involve communication from one bee to another, by claiming it may be a simple idiothetic movement that conveys no information. Conversely, experiments with robotic dummies were indeed able to induce some recruitment, which should not have been possible if the dance contains no information.

An article in the 18 September 2009 issue of New Scientist sets out evidence against the use by bees of the information in the dance.

The controversy persists, though it does so primarily due to an asymmetry between the two "camps"; those who study dance communication freely admit that odor is an essential component of the system, and even necessary at various stages of the recruitment process, including once a recruited forager reaches the vicinity of the resource (e.g.), while odor-plume advocates do not acknowledge that the dance contains any information whatsoever. Various experimental results demonstrate that the dance does convey information, but the use of this information may be context-dependent (e.g.), and this may explain why the results of earlier studies were inconsistent. In essence, both sides of the "controversy" agree that odor is used in recruitment to resources, but they differ strongly in opinion as to the information content of the dance.

Odor learning is usually tested by a method called the proboscis extension reflex.

Note: much of the research on the two competing hypotheses of communication has been restricted to Western honey bees (see the work of F.C. Dyer [3] though). Other species of *Apis* use variants on the same theme, and other types of bees use other methods altogether.

Trophallaxis

The exchange of food, trophallaxis, is also used by means to communicate and includes information on the quality of and thus competition for a food source, temperature and water demand, and the condition of the queen (Sebeok, 1990).

Primer Pheromones

Research that was published in November 2004, by scientists under the leadership of Dr. Zachary Huang, Michigan State University indicates that so called primer pheromones play an important part in how a honey bee colony adjusts its distribution of labourmost beneficially. In order to survive as a bee colony of sometimes 50,000 - 100,000 individual bees, the communal structure has to be adaptable to seasonal changes and the availability of food. The division of labourhas to adjust itself to the resources available from foraging. While the division of labourin a bee colony is quite complex, the work can be roughly seen as work inside the hive and outside the hive. Younger bees play a role inside the hive while older bees play a role outside the hive mostly as foragers. Huang's team found that forager bees gather and carry a chemical called ethyl oleate in the stomach. The forager bees feed this primer pheromone to the worker bees, and the chemical keeps them in a nurse bee state. The pheromone prevents the nurse bees from maturing too early to become forager bees. As forager bees die off, less of the ethyl oleate is available and nurse bees more quickly mature to become foragers. It appears that this control system is an example of decentralised decision making in the bee colony.

Cognition

Experiments by James Gould suggest that honey bees may have a cognitive map for information they have learned, and utilise it when communicating.

In one test reported in a 1983 issue of *Science News*, he moved a supply of sugar water 25% further away from a hive each day. The bees communicated to each other as usual on its location. Then he placed the sugar water on a boat anchored in the middle of a small lake. When scouts returned to the hive to communicate their find, other bees refused to go with them, not expecting to find food in the middle of a lake, even though they frequently flew over the lake to reach pollen sources on the opposite shore.

In another test related in the August 1986 issue of *Discover* ("A Honey of a Question: Are Bees Intelligent?"), Gould lured some bees to a dish of artificial nectar, then gradually moved it farther from the hive after they became accustomed to it. He marked the addicted bees, placed them in a darkened jar, and relocated them to a spot where the hive was still visible, but not the dish. When released one by one, the bees would appear disoriented for a few seconds, then fly directly for the covert dish. Seventy-three of 75 bees reached it in about 28 seconds. They apparently accomplished this feat by devising a new flight path based on a cognitive map of visible landmarks.

Bird songs - songbirds can be very articulate. African Grey Parrots are famous for their ability to mimic human language, and at least one specimen, Alex, appeared able to answer a number of simple questions about objects he was presented with. Parrots, hummingbirds and songbirds- display vocal learning patterns.

Chapter 9

Animal Mind

The question of animal minds asks whether it is meaningful to describe a non-human animal as having a mind.

Discussion of this subject is frequently confused by the fact that some schools of philosophy and psychology (*e.g.* radical behaviourism) would question whether one should ascribe mind to anyone else (or even to oneself). Such an approach would naturally deny the existence of animal minds.

However, even if one accepts that it is meaningful to talk about minds in one's own species, it remains an open question whether it is meaningful to talk about minds in other species.Though an animal "mind" features many of the same principals as a human mind, such as a basic sense of good and wrong, a sense of self protection, and a primitive ability to see anothers actions (good or bad)as a result of their own. In practice, the debate focuses on a few subjects within the field of animal cognition, particularly consciousness, self-concept, theory of mind, and tactical deception.

Cat Intelligence

Cat intelligence is the considered capacity of learning, thinking, problem solving, reasoning, and adaptability possessed by the domestic cat. Intelligence in cats is demonstrated by the capacity to develop and use tools, learn new behaviour techniques, apply previously acquired knowledge to new situations, communicate needs and desires within social groups, and respond to training cues. Mammalian neuroscientists have attempted to simulate the feline brain in order to understand the origins and operation of cat intelligence, but results to date are inconclusive and additional study is needed.

Brain Size and Surface Area

The brain size of the average cat is 5 centimeters in length and 30 grams. Since the average cat is 60 cm long and 3.3 kg, the brain makes up 1/12 of its length and 1/110 of its mass. Thus, the average cat's brain accounts for 0.9 percent of its total body mass, compared to 2 percent of total body mass in the average human. The surface area of a cat's cerebral cortex is approximately 83 cm^2. The modern human cerebral cortex is about 2500 cm^2. According to researchers at Tufts University School of Veterinary Medicine, the physical structure of human brains and that of cats are very similar; they have the same lobes in the cerebral cortex (the "seat" of intelligence) as humans do. Human brains also function the same way, conveying data via many identical neurotransmitters.

The Learning Cat

It is proven that cats learn by trial and error, observation and imitation. They can retain some information learned through exploration of their environment much longer than dogs. In one study, it was found that cats possess visual memory ability comparable to that of monkeys.

Intelligence by Breed

Ranking the intelligence of cats by breed is popular among pet owners, veterinarians and others, but the practice tends to run into difficulties. In general, the subject of cat intelligence rankings tends to be subjective. Cat breeder Norman Auspitz states the following: "As a rule, people seem to think the more active breeds have higher intelligence than the less active breeds. I will tell you that in feline agility, all breeds have done very well or very poorly as the case may be.. Having said that, there is no certified measure of cat intelligence and this general rule may be very anthropomorphic... until there is a credible definition of what might be meant by cat intelligence and a way to measure it, any comment anyone will make about the subject is, at best, speculation."

Observed Abilities of Cats

Inventing and Using a Tool: At least one cat was documented by a scientist to have adapted an object for use as a tool to add water to dry cat food—this tool-use being invented by the cat without any prior training by humans.

Opening Doors and Windows

Cats that are accustomed to being let outside, or that want to get into their home, may learn to open windows and doors. They are capable of learning different routes for entry and exit; for instance a cat might find the window in its owner's kitchen easier to open to exit the house, but to get in, they might have to use the screen door in the backyard. Also, they may learn to open cupboard doors to get to food. Cats' paws are not as effective at manipulation as human hands, owing to lack of an opposable thumb, but they can for instance learn to operate door lever handles by pulling them down, even though gripping the handle is difficult for cat paws.

Retrieving Items from Hard to Reach Places

A cat playing with a ball may suddenly find that the ball is under the couch. The cat will try different ways, changing paws, position, and other elements, the way a human would. This trial and error approach to puzzle solving can be demonstrated in the laboratory using Thorndike's puzzle boxes. In these boxes, cats must manipulate a series of levers in order to escape. They initially achieve this by trial and error, before committing the sequence to memory. They also use memory to reduce the amount of trial and error when encountering comparable novel situations e.g. new puzzle boxes. The cat may also be taught to get treats from high and hard to reach places, like on top of a refrigerator, or in a cupboard. Using the same logic as it did with the toy, the cat will get to each treat. A cat that has figured out where the cat food is kept may find that the food is inside a large bag. It might try to get in the bag or open it by means of removing the clip.

Using the Toilet

Because of their sensitive sense of smell, some cats prefer going outside to urinate and defecate, and rarely go in the same spot twice. Kittens are typically trained by their mothers to use a litter box and cover up their waste, so litter training rarely requires human intervention; once they understand where the litter box is, they will seek it out from then on. Cats can also be trained to make use of a toilet; some cats learn on their own after watching their owners, but for most cats, it is necessary to be taught by owners. In general, however, a toilet-trained cat is a rare animal, and successful toilet training depends both on the willingness of the animal to learn as well as on the patience of the owner to teach.

Playing Fetch

Some cats can be trained to play fetch with a varied degree of success (which is dependent on the cat and its mood). Siamese, Korats, Bengals and Burmese cats are well regarded as breeds that naturally carry objects in their mouths. They are easy to train to fetch and carry, again it may come naturally. Other breeds such as the Egyptian Mau, Maine Coon, Turkish Van, Savannah, Short Hair and Turkish Angora, and Bombay are also well known for an almost dog-like affinity for playing fetch. It is possible to get a cat to remain seated until an object is thrown. At that point, their sense of sight kicks in. As long as there is at least a remote chance of locating the thrown item, the cat will run off to find it. Once retrieved, waiting or a simple call is enough for the cat to return with the item (if it does not choose to do it themselves) and deposit it (usually) within arm's reach (or just outside as a possible form of dominance, making the owner change position). Chasing an object in the air is a natural cat hunting behaviour, and many cats will chase down a thrown toy for the sheer enjoyment of running and catching. Of course, any distraction and the cat may completely forget the game. This might suggest that a dog, which will do almost anything to please its owner due to pack instinct, will tend to focus on its game for both the owner and its pleasure, while a cat clearly plays the game for its own self-interest in chasing and pouncing. For a dog, the reward for a retrieved toy may be a vigorous petting or vocalisation of praise and also a second toss; for a cat, the only reward is typically the second throw.

Dog Intelligence

Dog intelligence is the ability of a dog to learn, think, and solve problems. Dog trainers, owners, and researchers have as much difficulty agreeing on a method for testing canine intelligence as they do for human intelligence. One specific difficulty is confusing a breed's genetic characteristics and a dog's obedience training with intelligence.

Certain breeds, like Border Collies, Poodles, German Shepherds, Shetland Sheepdogs, Rottweilers, Doberman Pinschers, Labrador Retrievers, Papillons, Australian Cattle Dogs and Golden Retrievers, are claimed by some to be "smarter" breeds of dogs because of their obedience. However, the ability and willingness to learn and obey commands is not the only possible measurement of intelligence. Other breeds, such as sled dogs and sight hounds, demonstrate intelligence in other ways.

Inherited Abilities

Dogs are pack animals. They understand social structure and obligations, and are capable of interacting with other members of the pack. Adult canines train their young by "correcting" them when they behave in an unacceptable manner (such as biting too hard or eating out of turn) and reward them for acceptable behaviour, by playing with them, feeding them, or cleaning them. They are also den animals. This means that they can easily learn behaviour related to keeping the den clean (such as housebreaking) and relaxing in an enclosed area (such as a crate during travel or for training).

Some breeds have been selectively bred for hundreds or thousands of years for the quality of learning quickly. That quality has been downplayed for other breeds in favourof other characteristics like the ability to track or hunt game, or to fight other animals. The capacity to learn basic obedience and complicated behaviour, however, is inherent in all dogs. Owners must simply be more patient with some breeds than with others. Nonetheless, inherited behaviour is not necessarily an indicator of intelligence. For example, a sheep herding breed, like a Border Collie, would be expected to learn how to herd sheep very quickly and might even perform the job with little training. The same breed, however, would be a challenge to train how to point and retrieve game. A Pointer often points to game instinctively and naturally retrieves game without damaging it, but training it to herd sheep would be difficult if not impossible.

Evaluation of Intelligence

The meaning of "intelligence" in general, not only in reference to dogs, is hard to define. Some tests measure problem-solving abilities and others test the ability to learn in comparison to others of the same age. Defining it for dogs is just as difficult. It is likely that dogs do not have the ability to premeditate an action to solve a problem. Some dogs may, however, have more drive to keep trying various things until they accidentally reach a solution and still others might have more ability to make the association between the "accident" and the result.

For example, the ability to learn quickly could be a sign of intelligence. It could be interpreted as a sign of blind subservience and a desire to please. In contrast, some dogs who do not learn very quickly may have other talents. An example is breeds that are not particularly interested in pleasing their owners, such as Siberian Huskies. Huskies are often fascinated with the myriad possibilities for escaping from yards and catching small animals, figuring out on their own numerous and often ingenious ways of doing both.

Assistance dogs are also required to be obedient at all times. This means they must learn a tremendous number of commands, understand how to act in a large variety of situations, and recognise threats to their human companions, some of which they might never before have encountered.

Many owners of livestock guardian breeds believe that breeds such as Great Pyrenees or the Kuvasz are not easily trained because their independent nature prevents them from seeing the point of such commands as "sit" or "down". The Molosser breeds are said to be particularly sensitive to physical or vocal aggression and, as such, are generally thought to respond to positive reinforcement-based methods of training. Hounds, (such as Beagles, Bloodhounds, and Basset Hounds), rank in the bottom tier of "The Intelligence of Dogs" list, but probably suffer from a certain approach to intelligence assessment. These dogs are bred to have a tenacious tracking mentality, taking advantage of their acute sense of smell, and less ability in "problem solving," which is the central task of Working and Herding dogs. In addition, many dog "authorities" are unaware of the Scenthound's extraordinary ability to perceive and evaluate things other than odors. They can detect pheromones, among other things, and may have the ability to evaluate a human's or another dog's personality or disposition from as far away as 300 feet. This can be described as "conditional intelligence" where the animal is quick to learn some things, while appearing reluctant to learn others.

Testing and Research

Certain intelligence tests involve the dog's ability to recognise and respond to a large vocabulary of commands. Other tests involve their desire or ability to respond to different situations. Just as with humans, there is a wide variety of interpretations as to what makes a dog "intelligent".

Various studies have attempted to rigorously classify intelligence of dogs. A recent example is animal psychologist Juliane Kaminski's paper in *Science* that demonstrated that Rico, a Border Collie, could learn over 200 words. Rico could remember the names of several items for up to four weeks after its last exposure (Kaminski eliminated the Clever Hans effect using strict protocols). Rico was also able to interpret phrases such as "fetch the sock" in terms of its component words (rather than considering its utterance to be a single word). Rico could also give the sock to a specified person. In 2008, Betsy, also a Border Collie, was featured on the cover of National Geographic Magazine. Betsy's intelligence rivaled that of Rico's in that she knew over 340 words and was able to correlate an object with a photographic image of the object, despite having seen neither before.

In his 1996 book *Good Natured*, ethologist Frans de Waal discusses an experiment on guilt and reprimands conducted on a female Siberian husky. The dog had the habit of shredding newspapers, and when her owner returned home to find the shredded papers and scold her she would act guilty. However, when the owner himself shredded the papers without the dog's knowledge, the dog "acted just as 'guilty' as when she herself had created the mess." De Waal concludes that the "guilt" displayed by dogs is not true guilt but rather the anticipation of the behaviour of an angry superior in a given situation.

A recent study in the journal PNAS concluded that dogs can feel complex emotions, like jealousy. Psychology research has shown that human faces are asymmetrical with the gaze instinctively moving to the right side of a face upon encountering other humans to obtain information about their emotions and state.

Research at the University of Lincoln (2008) shows that dogs share this instinct when meeting a human being, and only when meeting a human being (i.e., not other animals or other dogs). As such they are the only non-primate species known to do so.

School psychologist Kathy Coon developed the first intelligence test for dogs in 1976, with the work continuously revised through 2003. Assessments were developed to test short term memory, agility, ability to adapt, problem solving, unique detour problems, and to see how the dog reacts to conditions which he or she finds unacceptable. The performance of individual dogs was compared to over 100 dogs on which the test was standardised.

Additional breed norms were developed in her book, *The Dog Intelligence Test*. Stanley Coren ranked dog breeds by intelligence in his book *The Intelligence of Dogs* based on surveys done of dog trainers, the article for which contains a summary of the rankings obtained.

Cephalopod Intelligence

Cephalopod intelligence has an important comparative aspect in the understanding of intelligence because it relies on a nervous system fundamentally different from that of vertebrates. The cephalopod class of molluscs, particularly the Coleoidea subclass (cuttlefish, squid and octopuses), are considered the most intelligent invertebrates and an important example of advanced cognitive evolution in animals.

The scope of cephalopod intelligence is controversial, complicated by the challenges of studying these elusive and fundamentally different creatures. Classical conditioning of cephalopods has been reported, and one study (Fiorito and Scotto, 1992) even concluded that octopuses practise observational learning.

Figure: *An octopus in a zoo*

However, the latter idea is strongly disputed, and doubt has been shed on some other reported capabilities as well. In any case, impressive spatial learning capacity, navigational abilities, and predatory techniques remain beyond question.

Examples of Intelligence

Predation Techniques: Unlike most other molluscs, all cephalopods are active predators (with the possible exception of the bigfin squid). Their requirement to locate and capture their prey has been a probable driving force behind the development of their intelligence, uniquely advanced in their phylum.

The humboldt squid hunts schools of fish, showing extraordinary cooperation and communication in its hunting techniques. This is the first observation of such behaviour in invertebrates.

Crabs, the staple food source of most octopus species, present significant challenges with their powerful pincers and their potential to exhaust the cephalopod's respiration system from a prolonged pursuit. In the face of these challenges, octopuses will instead seek out lobster traps and steal the prize inside. They are also known to climb aboard fishing boats and hide in the containers that hold dead or dying crabs.

Dexterity

Dexterity, an ability essential for tool use and manipulation, is also found in cephalopods. The highly sensitive suction cups and prehensile arms of octopuses, squid, and cuttlefish are as effective at holding and manipulating objects as the human hand. However, unlike vertebrates, the motor skills of octopuses do not seem to depend upon mapping their body within their brains, as the ability to organise complex movements is not thought to be linked to particular arms.

One particularly clever octopus called Otto has been known to juggle his fellow tankmates around out of boredom, as well as throwing rocks and smashing the aquarium glass. On more than one occasion he even caused short circuits by crawling out of his tank and shooting a jet of water at the overhead lamp.

Communication

Another example of cephalopod intelligence is the communication that takes place between the more social species of squid. Some cephalopods are capable of rapid changes in skin colour and pattern through nervous control of chromatophores. This ability almost certainly evolved primarily for camouflage, but squids use colour, patterns, and flashing to communicate with one another in various courtship rituals. Caribbean Reef Squid can send one message via colour patterns to a squid on their right, while they send another message to a squid on their left.

Tool Use

Figure: *A small octopus (4-5 cm diameter) individual using a nut shell and clam shell as shelter.*

The octopus is one of the prime examples of an invertebrate animal which has repeatedly been shown to exhibit flexibility in its use of tools. At least four specimens of the Veined Octopus (*Amphioctopus marginatus*) have been witnessed retrieving discarded coconut shells, manipulating them, transporting them some distance, and then reassembling them to use as a shelter. This discovery was documented in the journal *Current Biology* and has been filmed. Most hermit crabs use discarded shells of other species for habitation and other crabs choose sea anemones to cultivate on their carapaces as camouflage; numerous insects use rocks, sand, leaves and so on as building materials, however none of this behaviour compares to the complexity of the octopus's fortress behaviour, which involves picking up and carrying a tool to use later on.

Fish Intelligence

According to Culum Brown from Macquarie University, "Fish are more intelligent than they appear. In many areas, such as memory, their cognitive powers match or exceed those of 'higher' vertebrates including non-human primates."

Memory

Fish capture by anglers has been shown to cause decreases in catchability. This therefore indicates that fish use their memory of negative experiences to associate capture with the stress response and therefore become less easy to catch.

This learning has been shown not just in carp but also paradise fish (*Macropodus opercularis*) which avoid places where they have experienced a single attack by a predator and continue to do so for many months. Also, several fish species are capable of learning complex spatial relationships and forming mental maps and integrate experiences which enable the fish to generate appropriate avoidance responses. This means that a fish can exhibit strong aversive behaviour if exposed to tissue damage or a predator. As a result, any reduction in the stressfulness of a capture by an angler, should be beneficial to angling in the long-term, since recapture of the fish should be less difficult.

Great Ape Language

Research into non-human great ape language has involved teaching chimpanzees, bonobos, gorillas, and orangutans to communicate with human beings and with each other using sign language, physical tokens, and lexigrams; see Yerkish. Some primatologists argue that the primates' use of these tools indicates their ability to use "language", although this is not consistent with some definitions of that term.

Questions in Animal Language Research

Animal language research attempts to answer the following questions:

- What problems can animals solve without language, and can they solve them better after they have received language training?
- Can the lessons learned in teaching animals be applied to human children?
- How, and how much, do animals' abilities to learn language differ from those of humans?
- Are the abilities that underlie language general or highly specialised?
- Where is the line between language and other forms of communication?

Apes that Demonstrate Understanding

Non-human animals have been recorded to have produced behaviours that are consistent with meanings accorded to human sentence productions. (A *production* is a stream of *lexemes* with semantic content. A language is grammar and a set of lexemes. A *sentence*, or statement, is a stream of lexemes that obeys a grammar, with a beginning and an end.) Some animals in the following species can be said to "understand" (*receive*), and some can "apply" (*produce*) consistent, appropriate, grammatical streams of communication. David Premack and Jacques Vauclair have cited language research for the following animals:

- Common Chimpanzees
- Bonobos
- Gorillas
- Orangutans.

Primate Use of Sign Language

Sign language and computer keyboards are used in primate language research because non-human primate vocal cords cannot close fully, and they have less control of the tongue and lower jaw. However, primates do possess the manual dexterity required for keyboard operation.

Many researchers into animal language have presented the results of the studies described below as evidence of linguistic abilities in animals. Many of their conclusions have been disputed.

It is now generally accepted that Apes can learn to sign and are able to communicate with humans. However, it is disputed as to whether they can form syntax to manipulate such signs.

Washoe

Washoe, a Common Chimpanzee, was caught in the wild in 1966. When she was about ten months old, she was received by the husband-and-wife research team of Beatrix T. Gardner and R. Allen Gardner. Chimpanzees are completely dependent until two years of age and semi-dependent until the age of four. Full adult growth is reached between 12 and 16 years of age. So the Gardners received her at a good age for research into language development. The Gardners tried to make Washoe's environment as similar as possible to what a human infant with deaf parents would experience. There was always a researcher or assistant in attendance during Washoe's waking hours. Every researcher communicated with Washoe by using American Sign Language (ASL), minimising the use of the spoken voice. The researchers acted as friends and companions to Washoe, using various games to make the learning as exciting as possible.

The Gardners used many different training methods:

- *Imitation:* After Washoe had learned a couple of words, she started, like chimpanzees usually do, to imitate naturally. For example, when she entered the Gardners' bathroom, she spontaneously made the sign for "toothbrush", simply because she saw one.
- *Babbling:* In this case, "babbling" does not mean vocal babbling. Instead, Washoe used untaught signs to express a desire. She used a begging gesture, which was not much different from the ASL signs "give me" and "come." (Human infants who are learning sign language often babble with their hands.)
- *Instrumental Conditioning:* The researchers used instrumental conditioning strategies with Washoe. For example, they taught the word "more" by using tickling as a reward. This technique was later applied to a variety of relevant situations.

The results of the Gardners' efforts were as follows:

- *Vocabulary:* When a sign was reported by three independent observers, it was added to a checklist. The sign had to occur in an appropriate context and without prompting. The checklist was used to record the frequency of a sign. A sign had to be used at least once a day for 15 consecutive days before it was deemed to have been acquired. Alternatively, a sign had to be used at least 15 days out of 30 consecutive days. By the end of the 22nd month of the project, thirty-four signs had been learned.

- *Differentiation:* Washoe used the sign "more" in many different situations until a more specific sign had been learned. At one point, she used the sign for "flower" to express the idea of "smell." After additional training, Washoe was eventually able to differentiate between "smell" and "flower."
- *Transfer:* Although the same object was presented for each learning trial (a specific hat, for example), Washoe was able to use the sign for other similar objects (e.g. other hats).
- *Combinations:* Washoe was able to combine two or three signs in an original way. For example, "open food drink" meant "open the fridge" and "please open hurry" meant "please open it quickly."

Nim Chimpsky

Linguistic critics challenged the animal trainers to demonstrate that Washoe was actually using language and not symbols. The *null hypothesis* was that the Gardners were using conditioning to teach the chimpanzee to use hand formations in certain contexts to create desirable outcomes, and that they had not learned the same linguistic rules that humans innately learn.

In response to this challenge, the chimpanzee Nim Chimpsky was taught to communicate using sign language in studies led by Herbert S. Terrace. In 44 months, Nim Chimpsky learned 125 signs. However, linguistic analysis of Nim's communications demonstrated that Nim's use was symbolic, and lacked grammar, or rules, of the kind that humans use in communicating via language. This constitutes a chimpanzee vocabulary learning rate of roughly 0.1 words per day. This rate is not comparable to the average college-educated English-speaking human who learns roughly 14 words per day between ages 2 and 22.

Sign Language in the Wild

A two-year study by researchers from the University of St Andrews of chimpanzees at Budongo Conservation Field Station in Uganda, concluded that this wild group make use of at least 66 distinct gestures to facilitate communication.

Plastic Tokens

Sarah and two other chimpanzees, Elizabeth and Peony, in the research programs of David Premack, demonstrated the ability to produce grammatical streams of token selections. The selections came from a vocabulary of several dozen plastic tokens; it took each of the chimpanzees hundreds of trials to reliably associate a token with a referent, such as an apple or banana. The tokens were chosen to be completely different in appearance from the referents. After learning these protocols, Sarah was then able to associate other tokens with consistent behaviours, such as negation, name-of, and if-then. The plastic tokens were placed on a magnetic slate, within a rectangular frame in a line. The tokens had to be selected and placed in a consistent order (a grammar) in order for the trainers to reward the chimpanzees.

One other chimpanzee, Gussie, was trained along with Sarah but failed to learn a single word. Other chimpanzees in the projects were not trained in the use of the tokens. All nine of the chimpanzees could understand gestures, such as supplication when asking for food; similarly, all nine could point to indicate some object, a gesture which is not seen in the wild. The supplication is seen in the wild, as a form of communication with other chimpanzees.

A juvenile Sumatran orangutan Aazk (named after the American Association of Zookeepers) who lived at the Roeding Park Zoo (Fresno, California) was taught by Gary L. Shapiro from 1973 to 1975 how to "read & write" with plastic children's letters, following the training techniques of David Premack. The technique of conditional discrimination was used such that the orangutan could eventually distinguish plastic letter (symbols) as representations of referents (e.g., object, actions) and "read" an increasingly longer series of symbols to obtain a referent (e.g., fruit) or "write" an increasingly longer series of symbols to request or describe a referent. While no claim of linguistic competence was made, Aazk's performance demonstrated design features of language, many similar to those demonstrated by Premack's chimpanzee, Sarah.

Kanzi

Kanzi, a Bonobo, is believed to understand more human language than any other nonhuman animal in the world. Kanzi apparently learned by eavesdropping on the keyboard lessons researcher Sue Savage-Rumbaugh was giving to her adoptive mother. Kanzi learned to communicate with a Lexigram board, pushing symbols that stand for words. The board is wired to a computer, so the word is then vocalised out loud by the computer. This helps Kanzi develop his vocabulary and enables him to communicate with researchers.

One day, Rumbaugh used the computer to say to Kanzi, "Can you make the dog bite the snake?" It is believed Kanzi had never heard this sentence before. In answering the question, Kanzi searched among the objects present until he found a toy dog and a toy snake, put the snake in the dog's mouth, and used his thumb and finger to close the dog's mouth over the snake. In 2001, Alexander Fiske-Harrison, writing in the *Financial Times*, observed that Kanzi was "asked by an invisible interrogator through head-phones (to avoid cueing) to identify 35 different items in 180 trials. His success rate was 93 per cent." In further testing, beginning when he was 7 ½ years old, Kanzi was asked 416 complex questions, responding correctly over 74% of the time. Kanzi has been observed verbalizing a meaningful noun to his sister.

Criticisms of Primate Language Research

Some scientists, including MIT linguist Noam Chomsky and cognitive scientist Steven Pinker, are skeptical about claims made for great ape language research. Among the reasons for skepticism are the differences in ease with which human beings and apes can learn language; there are also questions of whether there is a clear beginning and end to the signed gestures and whether the apes actually understand language or are simply doing a clever trick for a reward.

While vocabulary words from American Sign Language are used to train the apes, native users of ASL note that mere knowledge of ASL's vocabulary does not equate to ASL, but more closely reflects Pidgin Signed English which is not a full-fledged language. In the research involving Washoe, all researchers returned lists of signs Washoe used, with the exception of the one deaf native ASL user who reported no signs but many gestures. Native users of ASL make clear distinctions about what handshapes, palm orientations, and places of articulation signs must have to constitute linguistic activity. Signs must also be used combinatorially and in the correct grammatical sequence.

Thus, apes are seen as attempting to approximate these complex rules but are considered to be failing because of such malformations in the production of ASL signs. A precondition for a successful experiment with teaching a true sign language to primates ought to be ensuring that the main contact persons are all native speakers of the sign language, as it is otherwise analagous to trying to raise a human child as a speaker of a language to which you possess only a dictionary — with mispronunciations and worst of all giving only a pidgin model.

Great Ape Language in Fiction

Fantasy writer Edgar Rice Burroughs invented a fictitious great ape language called Mangani in his Tarzan books. This imagined language included such words as *Kreegah!* ("Beware!") and *Tarmangani* ("Great White Ape"). These words and others are sometimes used by cartoonists, and for facetious slang. (See Kreegah bundolo).

Writer Michael Crichton used the concept of great ape language in his 1980 novel *Congo*, in which a fictional gorilla named Amy communicates extensively with her keeper using ASL. This is also shown in the movie *Congo*.

The X-Files has an episode titled *Fearful Symmetry* in which a gorilla named Sophie communicates extensively with humans via sign language.

The 1987 film *Project X*, starring Matthew Broderick, had Chimpanzee use of American Sign Language as an important part of the plot.

The 2011 film *Rise of the Planet of the Apes* depicts apes using sign language to communicate with humans and with each other. Though the apes' brains have been enhanced in the film, there is an orangutan shown to use sign language prior to any enhancement.

Pigeon Intelligence

Pigeons have featured in numerous experiments in comparative psychology, including experiments concerned with animal cognition, and as a result we have considerable knowledge of pigeon intelligence.

Available data show, for example, that:

- Pigeons have the capacity to share attention between different dimensions of a stimulus, but (like humans and other animals) their performance with multiple dimensions is worse than with a single stimulus dimension.
- Pigeons can be taught relatively complex actions and response sequences, and can learn to make responses in different sequences.
- Pigeons readily learn to respond in the presence of one simple stimulus and withhold responding in the presence of a different stimulus, or to make different responses in the presence of different stimuli.
- Pigeons can discriminate between other individual pigeons, and can use the behaviour of another individual as a cue to tell them what response to make.
- Pigeons readily learn to make discriminative responses to different categories of stimuli, defined either by arbitrary rules (e.g. green triangles) or by human concepts (e.g. pictures of human beings).
- They do less well with categories defined by abstract logical relationships, e.g. "symmetrical" or "same", though some experimenters have successfully trained pigeons to discriminate such categories.

- Pigeons seem to require more information than humans for constructing a three-dimensional image from a plane representation.
- Pigeons seem to have difficulty in dealing with problems involving classes of classes. Thus they do not do very well with the isolation of a relationship among variables, as against a representation of a set of exemplars.
- Pigeons can remember large numbers of individual images for a long time, e.g. hundreds of images for periods of several years.

All these are capacities that are likely to be found in most mammal and bird species. In addition pigeons have unusual, perhaps unique, abilities to learn routes back to their home from long distances. This homing behaviour is different from that of birds that show migration, which usually occurs over a fixed route at fixed times of the year, whereas homing is more flexible; however similar mechanisms may be involved.

Additionally pigeons may be among the very few animals to pass the mirror test — which tests whether an animal recognises its reflection as an image of itself — along with common chimpanzees, bonobos, orangutans, dolphins, African grey parrots, crows, magpies, elephants, and humans.

Discrimination Abilities of Pigeons

In a famous article in 1995, Watanabe, Sakamoto and Wakita described an experiment which showed that pigeons can be trained to discriminate between paintings by Picasso and Monet.

The birds were first trained on a limited set of paintings: when the shown painting was a Picasso, the pigeon was able to obtain food by repeated pecking; when it was a Monet, pecking had no effect. After a while, the pigeons would only peck when shown Picasso paintings. They were then able to generalise, and correctly discriminate between paintings of the two painters not previously shown, and even between cubist and impressionist paintings (cubism and impressionism being the two stylistic schools Picasso and Monet belong to). When the Monet paintings were shown upside down, the pigeons were not able to properly categorise anymore; showing the cubist works upside down did not have such an effect.

In 1995, the authors won the humorous Ig Nobel Prize in psychology for this work.

In a later paper, Watanabe showed that if pigeons and human college students undergo the same training, their performance in distinguishing between Van Gogh and Chagall paintings is comparable.

Similar experiments had shown earlier that pigeons can be trained to distinguish between photos showing human beings and those that do not, and between photos showing trees and those that do not, among many other examples.

In all these cases, discrimination is quite easy for humans, even though the classes are so complex that no simple distinguishing algorithm or rule can be specified. It has therefore been argued that pigeons are able to form "concepts" or "categories" similar to humans, but that interpretation is controversial. Nevertheless, the experiments remain important and often cited examples in cognitive science.

Zoopharmacognosy

Zoopharmacognosy refers to the process by which non-Human animals self-medicate, by selecting and using plants, soils, and insects to treat and prevent disease. Coined by Dr. Eloy Rodriguez, a biochemist and professor at Cornell University, the word is derived from roots *zoo* ("animal"), *pharma* ("drug"), and *gnosy* ("knowing").

Observers have noticed that some species ingest non-foods, such as toxic plants, clay or charcoal, to ward off parasitic infestation or poisoning. Jane Goodall witnessed chimpanzees eating certain bushes to make themselves sick, and substantial evidence indicates that they swallow whole the leaves of certain rough-leaved plants, such as *Aneilema aequinoctialc*, in order to remove parasitic worms from their intestines. Some Brazilian parrots eat kaolin (a form of clay).

Illustrating the medicinal knowledge of some species, apes have been observed selecting a particular part of a medicinal plant by taking off leaves, then breaking the stem to suck out the juice. In an interview with Neil Campbell, Rodriguez describes the importance of biodiversity to medicine:

> *Some of the compounds we've identified by zoopharmacognosy kill parasitic worms, and some of these chemicals may be useful against tumors. There is no question that the templates for most drugs are in the natural world.*

William Astor has observed self-medicative behaviour in captive estrildid finches. The finches, of several different species, were observed to correctly target appropriate antibiotics out of a range made available to them, through which they were able to cure themselves of life threatening diseases (as opposed to relatively minor ailments like indigestion). These finches did not learn this self-medicative behaviour from their parents or others in the flock through social learning like apes and chimpanzees are suspected of doing. This ability of estrildid finches seems to be innate.

A female capuchin monkey in captivity was observed using tools covered in a sugar-based syrup to groom her wounds and those of her infant. Elephants in Africa will self-medicate by chewing on the leaves of a tree from the Boraginaceae family, which induces labor. Kenyans also use this tree for the same purpose.

Tool Use by Animals

Tools are used by some animals, particularly primates, to perform simple tasks such as the acquisition of food, or grooming. Originally thought to be a skill only possessed by humans, tool use requires some level of intelligence. Primates have been observed exploiting sticks and stones to accomplish tasks. Numerous bird species have also been noted as capable of using tools. The behaviour has also been observed in dolphins, elephants, otters, birds and octopuses. Tools may be used by animals for construction.

Opposable thumbs are a benefit in tool use, though creatures without hands have managed to use other body parts to their advantage, notably the mouth.

Types of Tools

Key to identifying tool use is defining what constitutes a tool. Researchers of animal behaviour have arrived at different formulations.

> *An object that has been modified to fit a purpose' or 'An inanimate object that one uses or modifies in some way to cause a change in the environment, thereby facilitating one's achievement of a target goal'.* —Hauser, 2000

> *the use of physical objects other than the animal's own body or appendages as a means to extend the physical influence realised by the animal —Jones and Kamil, 1973*

> *an object carried or maintained for future use —Finn, Tregenza, and Norman, 2009.*

The lack of a unique and sharp definition makes it difficult to identify many animal behaviours.

Uses

Tool use implies an animal has knowledge of the relationship between objects and their effects.

If an object is placed out of reach on a towel that itself is in reach, dogs, cats, and children will pull the towel to bring the object closer to them. But does this show knowledge about the nature of the world (declarative memory) or recall of rules already learnt (procedural)?

- Sticks can be used to break into termite nests for food or even to fight rivals. They are sometimes used for grooming.
- Stones can be used, again, to fight rivals. However, they may also be used to carve bits of wood by more intelligent animals.

Some species, such as the Woodpecker Finch of the Galapagos Islands, use particular tools as an essential part of their foraging behaviour. However, these behaviours are often quite inflexible and cannot be applied effectively in new situations. Several species have now been shown to be capable of more flexible tool use. A well known example is Jane Goodall's observation of chimpanzees "fishing" for termites in their natural environment, and captive great apes are often observed to use tools effectively; several species of corvids have also been trained to use tools in controlled experiments, or use bread crumbs for bait-fishing.

Tool Use by Specific Groups of Animals

Primates: The animals that make the widest use of tools are humans, who have developed mechanical, electric and electronic tools for multiple purposes, far in advance of even the most advanced non-human animals.

Research in 2007 shows that chimpanzees in the Fongoli savanna sharpen sticks to use as spears when hunting, considered the first evidence of systematic use of weapons in a species other than humans. Chimpanzees and bonobos were observed in the 1970s using sticks as probes to collect ants and termites. (Dr. Jane Goodall, on Nov. 4, 1960, observed a chimpanzee using a grass stalk to extract termites.) Also they have been observed cutting down the stick with their fingers and teeth so that it can fit into a hole in the ants' nest. They have even been observed using two tools, a stick to dig into the ant nest and a 'brush' made from grass stems with their teeth to collect the ants.

In West Africa chimpanzees have been observed banging nuts with a stone in order to crack them. Some troops use another stone whilst others use wooden clubs (heavy sticks). In one troop of chimpanzees it was observed that a female was using a stick to break into a bee hive to acquire honey. In an experiment a group of chimpanzees were presented with a model leopard with a moving head. There was soon commotion as leopards are one of the chimpanzees' predators. They were then observed clubbing the model with heavy quarterstaffs (fallen trees and/or branches). They continued doing this until the moving head had fallen off. Both bonobos and chimpanzees have also been observed making "sponges" out of leaves and moss that suck up water and are used as grooming tools.

Gorillas have been observed to use sticks to measure the depth of water and as "walking sticks" to support their posture when crossing deeper water (shown above).

Orangutans have also been observed to use sticks to measure the depth of water. It has also been observed that orangutans in Sumatra use sticks to acquire seeds from a certain fruit. This is because the lining of the inside of the fruit has hairs that sting. On the island of Kaja a male orangutan was observed using a pole to acquire fish from a net after observing local humans spear fishing.

Tool use has been observed in capuchin monkeys both in captivity and in their natural environments. In a captive environment, capuchins readily insert a stick into a tube containing viscous food that clings to the stick, which they then extract and lick. Capuchins also use a stick to push food from the center of a tube retrieving the food when it reaches the far end and as a rake to sweep objects or food toward themselves.

Wild capuchin monkeys in many areas use stone hammers and anvils to crack nuts and encased seed. They transport stones and nuts to an anvil for this purpose. Capuchins also use stones to excavate tubers and sticks to flush prey from inside rock crevices.

Birds

Many birds have been shown as capable of using tools. By Jones and Kamil's definition above, an Egyptian vulture dropping a bone on a rock would not be using a tool since the rock cannot be seen as an extension of the body. However the use of a rock manipulated using the beak to crack an ostrich egg would qualify the Egyptian vulture as a tool user. Many other species, including parrots, corvids and a range of passerines, have been noted as tool users.

New Caledonian Crows have been observed in the wild to use stick tools with their beaks to extract insects from logs. While young birds in the wild normally learn this technique from elders, a laboratory crow named "Betty" improvised a hooked tool from a wire with no prior experience. The Woodpecker Finch from the Galapagos Islands also uses simple stick tools to assist it in obtaining food. In captivity, a young Cactus Finch learned to imitate this behaviour by watching a Woodpecker Finch in an adjacent cage. Crows in urban Japan have innovated a technique to crack hard-shelled nuts by dropping them onto cross walks and letting them be run over and cracked by cars. They then retrieve the cracked nuts when the cars are stopped at the red light. In some towns in America, the crows would drop the walnuts onto busy streets and hope the cars would crack the nuts. Striated Herons (*Butorides striatus*) and Hooded Crows (*Corvus cornix*) use bait to catch fish.

Seagulls have been known to drop live oyster shells on paved and hard surfaces so that cars can drive over them and break the shell. So many get dropped that it is hard to drive down pavements safely near waterways. Certain species (e.g. the Herring Gull) have exhibited tool use behaviour, using pieces of bread as bait with which to catch goldfish, for example.

Common ravens are one of only a few species who make their own toys. They have been observed breaking off twigs to play with socially. Experimentation has also shown that ravens are capable of deducing how to pull up food suspended on a string.

Many owners of household parrots have observed their pets using various tools to scratch the back of their necks, including but not limited to discarded feathers, bottle caps, and popsicle sticks.

Cetaceans

As of 2005, scientists have observed limited groups of Bottlenose dolphins around the Australian Pacific using a basic tool. When searching for food on the sea floor, many of these dolphins were seen tearing off pieces of sponge and wrapping them around their "bottle nose" to prevent abrasions. Bottlenose dolphins have also been observed using conch shells to trap small fish, then lifting the shells out of the water causing the fish to fall into the dolphin's mouth.

Dolphins are often seen engaging in playful behaviour and create tools to use for entertainment. They have been observed to blow bubbles which they form into rings to play with. After creating the bubble ring, a dolphin will use its nose and body to maintain the shape of the bubble and keep it from floating to the surface.

Elephants

Elephants show a remarkable ability to use tools, despite having no hands. Instead, they use their trunk much like one would an arm. Elephants have been observed digging holes to drink water and then ripping bark from a tree, chewing it into the shape of a ball, filling in the hole and covering over it with sand to avoid evaporation. The elephant later went back to this spot for a drink. They also often use branches to swat flies or scratch themselves. Elephants have also been known to drop very large rocks onto an electric fence to either ruin the fence or cut off the electricity. Elephants in captivity have been observed moving objects from one location to another in order to use them as stools to stand upon, to access food that is otherwise out of reach.

Mustelids

Sea otters have been observed using stones to hammer abalone shells off the rocks. They hammer at a rate of 45 hits in 15 seconds or 180 rpm, and do it in two-three dives.

Veined Octopus

As of 2009, the octopus is the only invertebrate animal which has been conclusively shown to use tools. At least four specimens of the Veined Octopus (*Amphioctopus marginatus*) have been witnessed retrieving discarded coconut shells, manipulating them, transporting them some distance, stacking them and then reassembling them to use as a shelter.

This discovery was documented in the journal *Current Biology* and has been filmed on video. This can be seen as tool use, because it uses the shells as an extension of its skin for protection against predators.

Most hermit crabs use discarded shells of other species for habitation and other crabs choose sea anemones to cultivate on their carapaces as camouflage; numerous insects use rocks, sand, leaves and so on as building materials, however none of this is classified as tool use.

Fish

Several species of wrasses have been observed using rocks as anvils to crack bivalve shells.

Bibliography

Agamben, Giorgio: *The Open: Man and Animal,* Stanford University Press, UK, 2004.

Alexander, Shana: *The Astonishing Elephant,* Random House, New York, 2000.

Alger, Janet and Steven Alger: *Cat Culture: The Social World of a Cat Shelter,* Temple University Press, Philadelphia, 2003.

Allen, Karen Miller: *The Human-Animal Bond: An Annotated Bibliography,* Scarecrow Press, Metuchen, NJ, 1985.

Allister, Mark Christopher: *Eco-man: New Perspectives on Masculinity and Nature,* University of Virginia Press, Charlottesville, 2004.

Anderson, R. S.: *Pet Animals and Society,* Bailliere Tindall, London, 1984.

Baldick, Julian: *Animals and Shaman: Ancient Religions of Central Asia,* New York University Press, New York, 2000.

Baratay, Eric.: *Zoo: A History of Zoological Gardens in the West,* Reaktion, London, 2002.

Barton, M.: *Animal Rights,* Watts, London, 1987.

Bauston, Gene: *Battered Birds, Crated Herds: How We Treat the Animals We Eat,* Farm Sanctuary, Watkins Glen, NY, 1996.

Beck, Alan M. and Aaron Katcher: *Between Pets and People: The Importance of Animal Companionship,* Putnam, New York, 1983.

Bekoff, Marc and Jessica Pierce: *Wild Justice: The Moral Lives of Animals,* The University of Chicago Press, Chicago, 2009.

Bekoff, Marc: *The Smile of a Dolphin: Remarkable Accounts of Animal Emotions,* Discovery Books, New York, 2000.

Benton, Ted: *Natural Relations: Ecology, Animal Rights and Social Justice,* Verso, London, 1993.

Boehrer, Bruce: *A Cultural History of Animals in the Renaissance,* Oxford, UK: Berg Publishers, 2007.

Bright, M.: *Animal Language,* BBC Publications, London, 1984.

Brunner, Bernd: *Bears: A Brief History,* Yale University Press, New Haven, 2007.

Budiansky, S.: *The Covenant of the Wild: Why Animals Chose Domestication,* Morrow, New York, 1992.

Burt, Jonathan: *Rat,* Reaktion, London, 2004.

Calarco, Matthew, and Peter Atterton: *Animal Philosophy: Essential Readings in Continental Thought,* Continuum, London, 2004.

Carlson, Laurie: *Cattle: An Informal Social History,* Ivan R. Dee, Chicago, IL, 2001.

Carter, Paul: *Parrot,* Reaktion, London, 2005.

Chris, Cynthia: *Watching Wildlife,* University of Minnesota Press, Minneapolis, 2006.

Clark, S. and S. Lyster: *Animals and Their Moral Standing,* Routledge, London, 1997.

Clark, Stephen: *The Nature of the Beast.* Oxford University Press, New York, 1984.

Clutton-Brock, Juliet: *Domesticated Animals from Early Times,* University of Texas Press, Austin, TX, 1981.

Copeland, Marion W.: *Cockroach.* Reaktion, London, 2003.

Coren, Stanley: *The Pawprints of History: Dogs and the Course of Human Events,* Free Press, New York, 2002.

Crist, Eileen: *Images of Animals: Anthropomorphism and Animal Mind,* Temple University Press, Philadelphia, 1999.

DeGrazia, David: *Taking Animals Seriously: Mental Life and Moral Status,* Cambridge, New York, 1996.

Derr, Mark: *Dog's Best Friend: Annals of the Dog-human Relationship,* University of Chicago Press, Chicago, 2004.

Dolins, Francine: *Attitudes to Animals: Views on Animal Welfare,* Cambridge University Press, Cambridge, 1999.

Fine, Aubrey H.: *Handbook on Animal-Assisted Therapy: Theoretical Foundations and Guidelines for Practice,* Academic Press, UK, 2006.

Fitzgerald, Amy J.: *Animal Abuse and Family Violence: Researching the Interrelationships of Abusive Power,* Edwin Mellen Press, Lewiston, New York, 2005.

Fogle, Bruce: *Pets and Their People,* The Viking Press, New York, 1983.

Fox, Michael Allen: *The Case for Animal Experimentation,* University of California Press, Berkeley, CA, 1986.

Friend, Tim: *Animal Talk: Breaking the Codes of Animal Language,* Free Press, New York, 2004.

Fudge, Erica: *Brutal Reasoning: Animals, Rationality and Humanity in Early Modern England,* Cornell University Press, Ithaca, 2006.

Gallistel, C.R.: *Animal Cognition*, MIT Press, Cambridge, 1992.

Gates, P.: *Animal Communication,* Cambridge University Press, Cambridge, 1997.

Harbolt-Bosco, Tami: *Bridging the Bond: The Cultural Construction of the Shelter Pig,* Purdue University Press, 2003.

Harris, Marvin: *The Sacred Cow and the Abominable Pig: Riddles of Food and Culture,* Touchstone Books, New York, 1987.

Hearne, Vicki: *Animal Happiness,* HarperCollins, New York, 1994.

Hyland, Ann: *The Horse in the Ancient World,* Praeger, Westport, CT, 2003.

Jones, Susan D.: *Valuing Animals: Veterinarians and Their Patients in Modern America,* Johns Hopkins University Press, Baltimore, MD, 2003.

Katz, Jon: *The New Work of Dogs: Tending to Life, Love, and Family,* Villard, New York, 2003.

Kerasote, Ted: *Bloodties: Nature, Culture, and the Hunt,* Kodansha, New York, 1993.

Langley , Gill: *Animal Experimentation: The Consensus Changes,* Chapman and Hall, New York, 1989.

Lilequist , J.: *Forbidden History, the State, Society and the Regulation of Sexuality in Modern Europe,* Chicago University Press, Chicago, 1992.

MacDonald, Helen: *Falcon,* Reaktion, London, 2005.

Manning, Aubrey, and James Serpell: *Animals and Human Society,* Routledge, New York, 1994.

Mason, Jim and Peter Singer: *Animal Factories,* Crown Publishers, New York, 1980.

Masson, Jeffrey Moussaieff: *Dogs Never Lie About Love: Reflections on the Emotional World of Dogs,* Crown Publishers, New York, 1997.

McHugh, Susan: *Dog,* Reaktion, London, 2004.

Mitchell, W.J.T.: *The Last Dinosaur Book: The Life and Times of a Cultural Icon,* University of Chicago Press, Chicago and London, 1998.

Morris, Desmond: *The Naked Ape,* McGraw-Hill, New York, 1967.

Morton, Eugene S.: *Animal Talk: Science and the Voices of Nature,* Random House, New York, 1992.

Noske, Barbara: *Humans and Other Animals: Beyond the Boundaries of Anthropology,* Pluto Press, London, 1989.

Phillips, Ian: *Lost: Lost and Found Pet Posters from Around the World,* Princeton Architectural Press, New York, 2002.

Porter, V.: *Faithful Companions: The Alliance of Man and Dog,* Methuen, London, 1989.

Rifkin, Jeremy: *Beyond Beef: The Rise and Fall of the Cattle Culture,* Dutton, New York, 1992.

Ristau, C.: *Cognitive Ethology: The Minds of Other Animals*, Lawrence Erlbaum, Hillsdale, NJ, 1990.

Robbins, Louise E.: *Elephant Slaves and Pampered Parrots: Exotic Animals in Eighteenth-Century Paris*, Johns Hopkins University Press, Baltimore, 2002.

Roger, J.: *Buffon: A Life in Natural History,* Cornell University Press, Ithaca , NY, 1997.

Rogers, Katherine: *The Cat and the Human Imagination,* The University of Michigan Press, Ann Arbor, 2001.

Salem, Harry: *Animal Test Alternatives: Refinement, Reduction, Replacement,* M. Dekker, New York, 1995.

Sanders, Clinton: *Understanding Dogs,* Temple University Press, Philadelphia, 1999.

Sax, Boria: *Crow,* Reaktion Books, London, 2003.

Sharpe, Robert: *Science on Trial: The Human Cost of Animal Experiments,* Awareness Books, Sheffield, UK, 1994.

Sleigh, Charlotte: *Ant,* Reaktion, London, 2003.

Stott, Rebecca: *Oyster,* Reaktion, London, 2004.

Stutesman, Drake: *Snake,* Reaktion, London, 2005.

Swan, James: *In Defense of Hunting*, HarperCollins, New York, 1995.

Tabor, Roger K.: *The Wild Life of the Domestic Cat,* Arrow Books, London, 1983.

Taylor, R.E. and R. Bogart: *Scientific Farm Animal Production: An Introduction to Animal Science*, MacMillan, New York, 1988.

Thomas, Elizabeth Marshall: *The Tribe of the Tiger,* Simon and Schuster, New York, 1994.

Topsells, E.: *History of Four-footed Beast,* Nelson Hall, Chicago, 1981.

Ucko, Peter J. and G.W. Dimbleby: *The Domestication and Exploitation of Plants and Animals*, Aldine Publishing Company, Chicago, IL, 1969.

Willis, R.G.: *Signifying Animals: Human Meaning in the Natural World,* Unwin Hyman, London, 1990.

Yarri, Donna: *The Ethics of Animal Experimentation: A Critical Analysis And Constructive Christian Proposal,* Oxford University Press, Oxford, 2005.

Zeuner, Frederick E.: *A History of Domesticated Animals*, Hutchinson, London, 1963.

Index

❑❑❑